Earth & Space Series

Earth, Weather, & Waste

— Grade 3 —

Written by Tracy Bellaire

The activities in this book have two intentions: to teach concepts related to earth and space science and to provide students the opportunity to apply necessary skills needed for mastery of science and technology curriculum objectives.

The experiments in this book fall under twelve topics that relate to three aspects of earth and space science: **Exploring Soils in the Environment, Rocks and Minerals,** and **Stars and Planets**. In each section you will find teacher notes designed to provide you guidance with the learning intention, the success criteria, materials needed, a lesson outline, as well as provide some insight on what results to expect when the experiments are conducted. Suggestions for differentiation are also included so that all students can be successful in the learning environment.

Tracy Bellaire is an experienced teacher who continues to be involved in various levels of education in her role as Differentiated Learning Resource Teacher in an elementary school in Ontario. She enjoys creating educational materials for all types of learners, and providing tools for teachers to further develop their skill set in the classroom. She hopes that these lessons help all to discover their love of science!

Published in Canada by:
On The Mark Press
15 Dairy Avenue, Napanee, Ontario, K7R 1M4
www.onthemarkpress.com

Funded by the
Government
of Canada

Canadä

OTM2154   ISBN: 9781770789609
© On The Mark Press

At A Glance

Learning Intentions

Learning Intentions	Types of Soil	Soils in Your Environment	Growing in the Garden	Living in the Soil	At Work in the Soil	Erosion	Rock Types	Exploring Minerals	Fun with Rocks	Our Solar System	The View from Earth	Constellations
Knowledge and Understanding Content												
Identify the four components of soil and explore the different types of soil using the senses	•											
Identify and describe the components that are in soils by exploring sieving and sedimentation techniques		•										
Recognize and describe how different types of soil affect the growth of plants			•									
Describe how living things are interdependent with the soil they live in				•								
Describe the benefit of earthworms in making nutrient rich soil, and how composting provides nutrients to soil					•							
Determine the effects of water on soil						•						
Describe the three different rock types and discover the types that are in the neighborhood							•					
Identify minerals and describe them according to their properties; conduct a rock study								•				
Determine the presence of carbonates, and how rocks become eroded									•			
Identify and describe the unique features of each planet in our solar system										•		
Describe and demonstrate the positions of the sun, Earth, and its moon in our solar system											•	
Identify the constellations in our night sky; retell the myth behind a constellation												•
Thinking Skills and Investigation Process												
Make predictions, formulate questions, and plan an investigation			•	•	•	•			•		•	
Gather and record observations and findings using drawings, tables, written descriptions	•	•	•	•	•	•	•	•	•	•	•	•
Recognize and apply safety procedures in the classroom	•	•	•	•	•	•	•	•	•	•	•	•
Communication												
Communicate the procedure and conclusions of investigations using demonstrations, drawings, and oral or written descriptions, with use of science and technology vocabulary	•	•	•	•	•	•	•	•	•	•	•	•
Application of Knowledge and Skills to Society and the Environment												
Assess ways humans can positively affect the quality of soils in the environment					•	•						

2

OTM2154   ISBN: 9781770789609
© On The Mark Press

TABLE OF CONTENTS

OTM2154   ISBN: 9781770789609
© On The Mark Press

Teacher Assessment Rubric

Student's Name: _____   Date: _____

Success Criteria	Level 1	Level 2	Level 3	Level 4
Knowledge and Understanding Content				
Demonstrate an understanding of the concepts, ideas, terminology definitions, procedures and the safe use of equipment and materials	Demonstrates limited knowledge and understanding of the content	Demonstrates some knowledge and understanding of the content	Demonstrates considerable knowledge and understanding of the content	Demonstrates thorough knowledge and understanding of the content
Thinking Skills and Investigation Process				
Develop hypothesis, formulate questions, select strategies, plan an investigation	Uses planning and critical thinking skills with limited effectiveness	Uses planning and critical thinking skills with some effectiveness	Uses planning and critical thinking skills with considerable effectiveness	Uses planning and critical thinking skills with a high degree of effectiveness
Gather and record data, and make observations, using safety equipment	Uses investigative processing skills with limited effectiveness	Uses investigative processing skills with some effectiveness	Uses investigative processing skills with considerable effectiveness	Uses investigative processing skills with a high degree of effectiveness
Communication				
Organize and communicate ideas and information in oral, visual, and/or written forms	Organizes and communicates ideas and information with limited effectiveness	Organizes and communicates ideas and information with some effectiveness	Organizes and communicates ideas and information with considerable effectiveness	Organizes and communicates ideas and information with a high degree of effectiveness
Use science and technology vocabulary in the communication of ideas and information	Uses vocabulary and terminology with limited effectiveness	Uses vocabulary and terminology with some effectiveness	Uses vocabulary and terminology with considerable effectiveness	Uses vocabulary and terminology with a high degree of effectiveness
Application of Knowledge and Skills to Society and Environment				
Apply knowledge and skills to make connections between science and technology to society and the environment	Makes connections with limited effectiveness	Makes connections with some effectiveness	Makes connections with considerable effectiveness	Makes connections with a high degree of effectiveness
Propose action plans to address problems relating to science and technology, society, and environment	Proposes action plans with limited effectiveness	Proposes action plans with some effectiveness	Proposes action plans with considerable effectiveness	Proposes action plans with a high degree of effectiveness

OTM2154   ISBN: 9781770789609
© On The Mark Press

Student Self Assessment Rubric

Name: _____ Date: _____

Put a check mark ✔ in the box that best describes you.

Expectations	Always	Almost Always	Sometimes	Needs Improvement
I am a good listener.				
I followed the directions.				
I stayed on task and finished on time.				
I remembered safety.				
My writing is neat.				
My pictures are neat and colored.				
I reported the results of my experiment.				
I discussed the results of my experiment.				
I know what I am good at.				
I know what I need to work on.				

1. I liked _____

_____

2. I learned _____

_____

3. I want to learn more about _____

_____

INTRODUCTION

The activities in this book have two intentions: to teach concepts related to earth and space science and to provide students the opportunity to apply necessary skills needed for mastery of science and technology curriculum objectives.

Throughout the experiments, the scientific method is used. The scientific method is an investigative process which follows five steps to guide students to discover if evidence supports a hypothesis.

1. **Consider a question to investigate.**
   For each experiment, a question is provided for students to consider. For example, "Which soil type is best for growing plants?"

2. **Predict what you think will happen.**
   A hypothesis is an educated guess about the answer to the question being investigated. For example, "I believe that a mixture of loam, sand, and clay is the best for growing plants". A group discussion is ideal at this point.

3. **Create a plan or procedure to investigate the hypothesis.**
   The plan will include a list of materials and a list of steps to follow. It forms the "experiment".

4. **Record all the observations of the investigation.**
   Results may be recorded in written, table, or picture form.

5. **Draw a conclusion.**
   Do the results support the hypothesis? Encourage students to share their conclusions with their classmates, or in a large group discussion format.

The experiments in this book fall under twelve topics that relate to three aspects of earth and space science: **Exploring Soils in the Environment, Rocks and Minerals,** and **Stars and Planets**. In each section you will find teacher notes designed to provide you guidance with the learning intention, the success criteria, materials needed, a lesson outline, as well as provide some insight on what results to expect when the experiments are conducted. Suggestions for differentiation are also included so that all students can be successful in the learning environment.

ASSESSMENT AND EVALUATION:

Students can complete the Student Self-Assessment Rubric in order to determine their own strengths and areas for improvement. Assessment can be determined by observation of student participation in the investigation process. The classroom teacher can refer to the Teacher Assessment Rubric and complete it for each student to determine if the success criteria outlined in the lesson plan has been achieved. Determining an overall level of success for evaluation purposes can be done by viewing each student's rubric to see what level of achievement predominantly appears throughout the rubric.

OTM2154   ISBN: 9781770789609
© On The Mark Press

TYPES OF SOILS

LEARNING INTENTION:

Students will learn about the four components of soil and they will explore the different types of soil using their senses.

SUCCESS CRITERIA:

- identify the four components of soil
- identify different types of soil
- describe different types of soil using senses
- gather data and record observations in a chart
- make conclusions about the differences in soils
- make a connection about soil and its importance in the environment

MATERIALS NEEDED:

- a copy of "What is Soil?" Worksheet 1 for each student
- a copy of "Types of Soil" Worksheet 2 for each student
- a copy of "Reviewing the Facts!" Worksheet 3 for each student
- a copy of "Exploring the Soils" Worksheet 4, 5, 6, 7, and 8 for each student
- samples of soils such as sand, clay, silt, loam, and topsoil (enough for 4 or 5 groups)
- magnifying glasses, toothpicks, newspaper, clipboards, pencils, chart paper, markers

PROCEDURE:

*This lesson can be done as one long lesson, or be divided into two shorter lessons.

1. Give students Worksheet 1 and 2. Read through the information with them about the components of soil and about the different types of soil to ensure their understanding.

2. Have students engage in a 'Turn and Talk' activity where they turn to a partner and discuss the information that they have just learned about in step 1. Some guiding discussion questions that students can use:

   - What is soil made up of?
   - What are some different types of soil?
   - What have you learned about sand? Clay? Silt? Loam? Topsoil?

3. Give students Worksheet 3 to complete.

4. Explain to students that they will do some exploring with the different soil types. Give them Worksheet 4. Read through the question, materials needed, and what to do sections with them to ensure their understanding. Divide students into 4 or 5 groups. Send each group to a soil station in the room, where there are samples of each of the soil types. Give students Worksheets 5, 6, 7, and 8, and the materials to conduct the exploration. Using their senses of sight, smell, and touch, students will make and record observations of the soils, make conclusions and connections.

DIFFERENTIATION:

Slower learners may benefit by working as a small group with teacher direction and support in order to produce satisfactory written descriptions of their findings while exploring with the soils.

For enrichment, faster learners could write about a situation or place where they think that each of the soil types would be useful in our environment.

What is Soil?

Soil is made up of four things. These four things are **tiny rocks, water, air,** and **humus.** Let's read more about how soil is made!

When large rocks experience changes in temperature, they often crack into smaller pieces of rock. When wind and water hit against rocks over time, this also can make rocks crack into smaller pieces. These tiny rocks and minerals become part of soil.

Soil has spaces in it. In these spaces are water and air.

Humus is made when rotted plants and animal matter break down over time into smaller pieces. The humus mixes with tiny rocks, water, and air to make soil. Humus provides nutrients and helps the soil to keep moist.

OTM2154 ISBN: 9781770789609
© On The Mark Press

Types of Soil

There are different types of soils. Types of soil are **sand, clay, silt, loam,** and **topsoil.** Let's read more about these soils!

Topsoil is made up of sandy soil, clay, and humus. It holds water very well. The humus in it makes it rich in nutrients.

Loam is made up of sandy soil, clay, and silt. Loam is good for growing plants because it stays moist. Water and air can move and flow through its particles.

Sand is made up of tiny rocks and mineral pieces. The particles in sandy soil are bigger and loose. Sand does not hold water or nutrients very well. Most plants do not grow well in this soil.

Silt is made up of sandy soil and clay. It holds water well, but can be easily blown away by wind and washed away by water.

Clay has very fine rock particles in it. The rock grains that make up clay are small and close together. Clay holds water well, but there is not much air in it for plants to grow well.

OTM2154   ISBN: 9781770789609
© On The Mark Press

Reviewing the Facts!

Let's Review

Complete the diagram by naming the four things that make up soil.

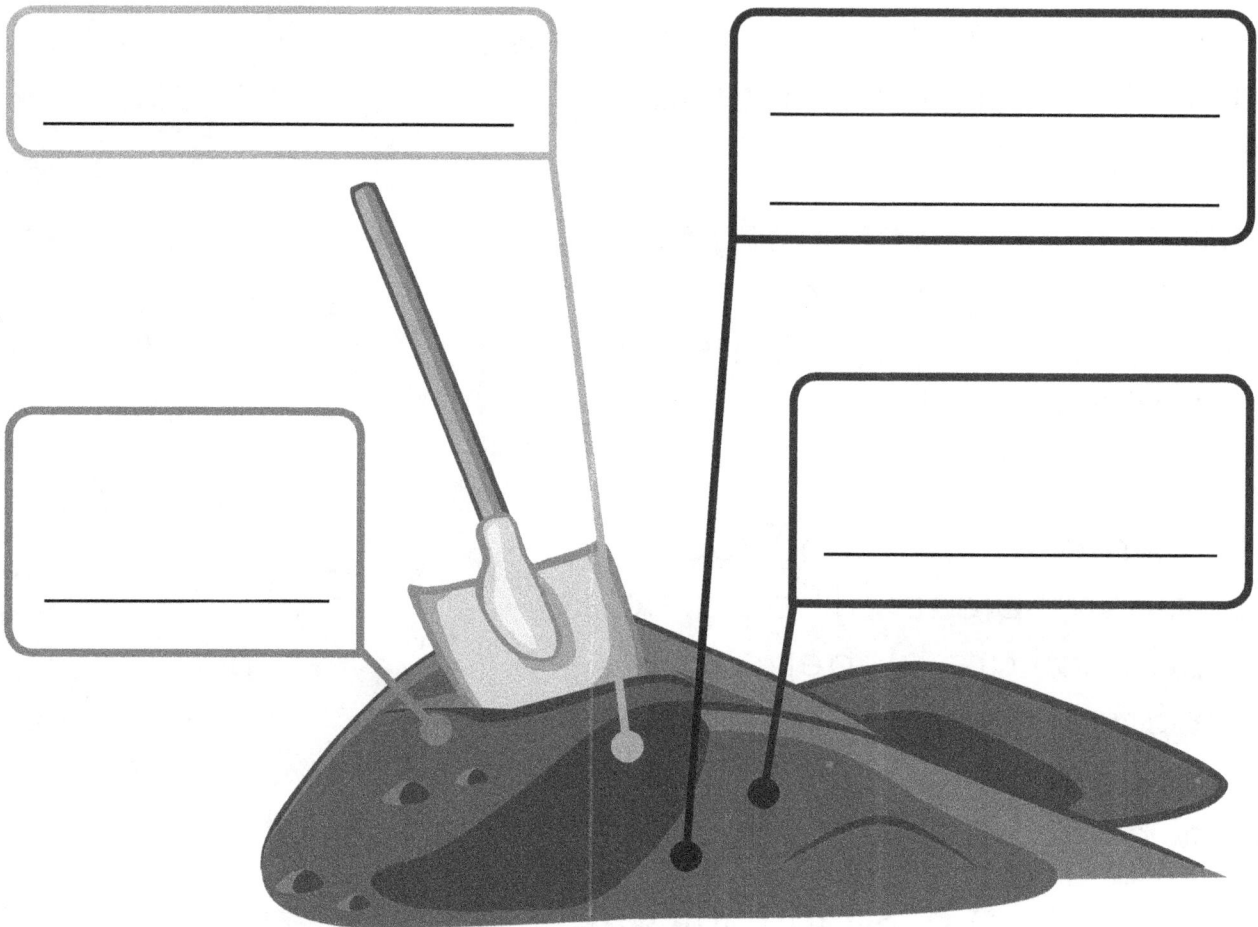

_____

_____

_____

_____

List the types of soil that you learned about:

_____

_____

_____

_____

_____

Which soil would you put in your garden to grow some plants?

_____

OTM2154   ISBN: 9781770789609
© On The Mark Press

Exploring the Soils

Question: How would you describe the different types of soils?

You'll need:

- newspaper
- toothpicks
- a magnifying glass

- different types of soils such as sand, clay, silt, loam, and topsoil

What to do:

1. Using the magnifying glass, some toothpicks, and your sense of sight, describe how each of the soils look. Record your answers in the chart on Worksheet 5.

2. Use your sense of touch to describe how each of the soils feel. Record your answers in the chart on Worksheet 6.

3. Use your sense of smell to describe how each of the soils smell. Record your answer on Worksheet 7.

4. On Worksheet 7, illustrate what each of the soils look like. Label them.

5. Make some conclusions about soil. Record them on Worksheet 8.

6. Make some connections about soil and our environment by answering the challenge question on Worksheet 8.

Name:

Let's Explore

Use your sense of sight to describe how each of the soils look.

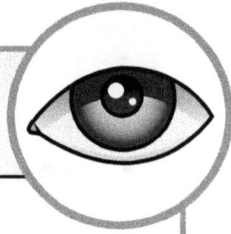

Soils	Describe what it looks like
Sand	
Clay	
Silt	
Loam	
Topsoil	

OTM2154   ISBN: 9781770789609
© On The Mark Press

Name:

Use your sense of touch to describe how each of the soils feel.

Soils	Describe what it feels like
Sand	
Clay	
Silt	
Loam	
Topsoil	

OTM2154   ISBN: 9781770789609
© On The Mark Press

Use your sense of smell to describe how each of the soils smell.

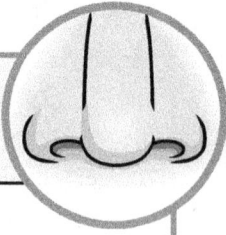

Soils	Describe what it smells like
Sand	
Clay	
Silt	
Loam	
Topsoil	

Draw a detailed illustration of each of the soils.

OTM2154   ISBN: 9781770789609
© On The Mark Press

Name:

Let's Conclude

What have you discovered about soil? What conclusions can you make?

_____

_____

_____

_____

What do you **wonder** about soil?

_____

_____

_____

Let's Connect It!

Challenge Question:

Why do you think soil is so important to our environment?

_____

_____

_____

_____

_____

_____

_____

_____

SOILS IN YOUR ENVIRONMENT

LEARNING INTENTION:

Students will learn about the components that are in their neighborhood soils by exploring sieving and sedimentation techniques.

SUCCESS CRITERIA:

- collect four different samples of soil
- illustrate and describe the origin of four different samples of soil
- describe samples of soil using the senses
- use a sieving technique to determine coarse, medium, and fine soil
- use a sedimentation technique to determine the components in a soil sample
- make and record observations using illustrations and words
- make a conclusion about the components in a soil sample
- make a connection about soil and its usefulness in the environment and to living things

MATERIALS NEEDED:

- a copy of "Going on a Soil Hunt!" worksheet 1 and 2 for each student
- a copy of "Soil Sleuthing!" worksheet 3, 4, 5, 6, and 7 for each student
- students will collect four different samples of soils from the neighborhood such as schoolyard soil, playground soil, garden/flower bed soil, grassy field soil, lawn soil, beach/riverbed soil, forest soil, or marsh soil
- small garden shovels (one per group)
- medium sized zip-lock bags (four per group)
- wide mesh sieves (one or two per group)
- fine mesh sieves (one or two per group)
- glass jars with lids (four per group)
- a jug of water (one per group)
- magnifying glasses, toothpicks, newspaper, clipboards, pencils, chart paper, markers
- *iPods, iPads, cameras (optional)*

PROCEDURE:

*This lesson can be done as one long lesson, or be divided into four shorter lessons.

1. Explain to students that they will explore the components that are in their neighborhood soils. They will take a walk in the neighborhood to collect four soil samples, which they will use to examine what is in them. At this point, divide students into groups of four, and have each person in the group collect a different soil sample, from a different location. Give each group a small garden shovel and four zip-lock bags. Give each student a clipboard and pencil, and Worksheets 1 and 2. Instruct students to draw a detailed picture of the soil sample they are taking, as they collect it. Then they will provide a description of where they are taking it from. An option is to have students take photos of where they are taking the soil sample from, using an iPod, iPad, or camera.

2. Create soil stations in the classroom where students can explore the components of their samples they collected. Students who were working in a group of four to collect soil samples should be seated together at the same soil station in the classroom. Give them Worksheet 3. Read through the question, materials needed, and what to do sections with them to ensure their understanding. Give each student in the group Worksheet 4 and the materials to begin the first part of the exploration using the toothpicks and magnifying glasses. **After completing Worksheet 4, instruct students to share and compare with all members of their group what they have discovered about the soil sample they collected.**

3. Give students Worksheet 5. Read through the information with the students to ensure their understanding of the next task. It may also be beneficial to discuss the meaning of 'coarse, medium, and fine', in terms of soil consistency.

OTM2154   ISBN: 9781770789609
© On The Mark Press

Give the sieves to the groups so that they can continue on to the next part of their soil exploration. **After completing Worksheet 5, instruct students to share and compare with all members of their group what they have discovered about the soil sample they collected.**

4. Give students Worksheet 6. Read through the information with the students to ensure their understanding of the next task. It may also be beneficial to discuss the meaning of 'soil components'. Give the jars and water jugs to the groups so that they can continue on to the next part of their soil exploration. An option is to have students use the iPods, iPads, or a camera to take a "before sedimentation" picture and an "after sedimentation" picture. **After completing Worksheet 7, instruct students to share and compare with all members of their group what they have discovered about the soil sample they collected.**

5. A follow-up option is to come back together as a large group, and have one member from each small group present the findings of one soil collection type to the large group. This could promote discussion about the benefits of different soil types in their neighborhood.

DIFFERENTIATION:

Slower learners may benefit by working as a small group with teacher direction in order to collect and investigate the components of their soils. This support will help to produce satisfactory written descriptions of their findings while exploring with the soils.

For enrichment, faster learners could write a short summary of what they have learned about the components in soils, then attach this to their before and after pictures that they took in step 4 of the procedure section. This would make for an interesting bulletin board display!

Going on a Soil Hunt!

You have learned about different types of soil. Now let's go on a hunt to explore what soils are in **your** neighborhood grounds!

In the boxes below, illustrate the soil samples that your group collects and describe exactly where you took each from.

This soil sample was taken from: _____

_____

_____

_____

_____

_____

_____

This soil sample was taken from: _____

_____

_____

_____

_____

_____

_____

OTM2154   ISBN: 9781770789609
© On The Mark Press

Name: _____

Continue to illustrate the soil samples that your group collects and describe exactly where you took each from.

This soil sample was taken from: _____

_____

_____

_____

_____

_____

_____

This soil sample was taken from: _____

_____

_____

_____

_____

_____

_____

OTM2154   ISBN: 9781770789609
© On The Mark Press

Soil Sleuthing!

Question: What soils are in the samples you found in your environment?

You'll need:

- newspaper
- toothpicks
- a jug of water
- 4 glass jars with lids
- a fine mesh sieve
- a wide mesh sieve
- 4 magnifying glasses
- 4 soil samples collected in your neighborhood

What to do:

1. Spread the newspaper on the table. Then pour out **only some** of your soil sample onto it.

2. Using the magnifying glasses, some toothpicks, and your senses, each member of the group will describe the soil sample that he/she collected. Record your answers on Worksheet 4. *(Each group member will have a soil sample and Worksheet 4)*. Share and compare your findings.

3. Each group member will continue investigating what is in their soil sample by using the sieves and the instructions on Worksheet 5.

4. After each group member has sieved and completed Worksheet 5, share and compare your findings about each of your soil samples.

5. Each group member will continue investigating what is in their soil sample by using the glass jars, water, and instructions on Worksheet 6. Make and record your observations by illustrating the layers in your jar.

6. Make some conclusions and connections about soil in your environment on Worksheet 7.

OTM2154 ISBN: 9781770789609
© On The Mark Press

Name:

Let's Examine!

Use your senses to give a detailed description of a soil sample that you collected in your neighborhood.

Soil Sample # _____	
Where did you find it?	
What does it look like?	
What does it feel like?	
What does it smell like?	

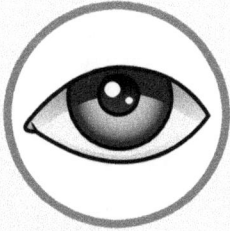

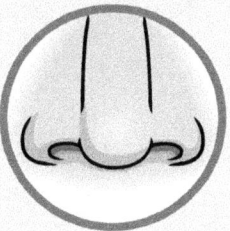

| Name:

Use the sieves to investigate what is in your soil sample.

- shake **some** of your soil sample through the wide mesh sieve onto a newspaper
- observe what is left in the sieve, observe what is on the newspaper
- put the soil that is on the newspaper into a fine mesh sieve, shake it onto another newspaper
- observe what is on this newspaper

Soil Sample # _____

Use pictures and words to tell what is left in the wide mesh sieve.	This is **coarse** soil.
Use pictures and words to tell what is on the first newspaper.	This is **medium** soil.
Use pictures and words to tell what is on the second newspaper.	This is **fine** soil.

OTM2154 ISBN: 9781770789609
© On The Mark Press

Did you Know?

When soil and water are blended together then left to settle, the soil will separate, and settle into layers. This is called **sedimentation**.

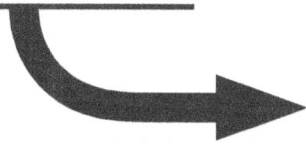

WATER
FINE CLAY
SILT
SAND
GRAVEL

Let's Try It!

Use a glass jar and water to investigate what is in your soil sample.

- fill a glass jar half full of a sample of soil
- pour water into the jar until it is about 3/4 full
- put the lid on the jar tightly
- shake it until the water and soil are blended
- let the jar stand for a day
- then make and record your observations on Worksheet 7

OTM2154 ISBN: 9781770789609
© On The Mark Press

Illustrate what the soil in the jar looked like a day later. Label the layers.

This was the soil sample taken from _____

_____

Let's Conclude

What is the main component of your soil sample?

_____

_____

_____

Challenge Question:

Explain why this type of soil might be helpful to living things in and near it.

_____

_____

_____

OTM2154   ISBN: 9781770789609
© On The Mark Press

GROWING IN THE GARDEN

LEARNING INTENTION:

Students will learn about how different types of soil affect the growth of plants.

SUCCESS CRITERIA:

- plant seeds in different soil types
- make and record a prediction about the soil type that will produce the healthiest plant
- water and measure plant growth of each plant every other day
- gather data and record observations in a chart, using diagrams and measurements
- create a bar graph to display the results of the plant growth in each soil type
- make a conclusion about the soil type that best produces the healthiest plants

MATERIALS NEEDED:

- a copy of "How Does Your Garden Grow?" Worksheet 1, 2, 3, 4, and 5 for each student
- small planting containers (3 per student, or 3 per pair of students)
- soils such as sand, clay, and loam
- a few packets of seeds such as pea, bean, or radish
- about 6 measuring cups, about 6 small garden shovels
- newspaper, rulers, masking tape, a few measuring cups
- access to water and sunshine
- clipboards, pencils, chart paper, markers, centimetre grid paper

PROCEDURE:

*This lesson will span over a three week period.

1. Students will experiment with plant growth in different soil types. Give them Worksheet 1. Read through the question, materials needed, and what to do sections with them to ensure their understanding. *Divide students into pairs or have them work individually.* Set up a few soil stations in the room, where there are samples of each of the soil types. Give students Worksheet 2 and the materials to begin the experiment. Students will record predictions about which soil will grow the healthiest plant. They will plant seeds in containers of soils, and place them in a warm, sunny place in the room.

2. Students are to water their plants every other day, giving each the same amount of water. Using Worksheets 2, 3, and 4, students will make and record observations of plant growth in each of the soil types over a three week period.

3. Students will display their data in a bar graph on Worksheet 5. The bar graph can be created either vertically or horizontally. Students will make a conclusion about what soil type is best to use for growing plants.

DIFFERENTIATION:

Slower learners may benefit by working as a small group with teacher direction to plant seeds in each soil type. Then, accurately measure the amount of water given to each plant, as well as measure and record the growth of each plant, using Worksheets 2, 3, and 4. This would result in one set of data to graph, which could be done together, using chart paper and markers.

For enrichment, faster learners could create a diagram of a garden on centimetre grid paper. Criteria to meet could be:

- estimate, measure, and record the length of each side of your garden
- calculate the perimeter of your garden
- calculate the area of your garden
- using pictures and words, tell about what you would grow in your garden

OTM2154   ISBN: 9781770789609
© On The Mark Press

How Does Your Garden Grow?

Question: What soil type is best for growing plants?

You'll need:

- masking tape
- newspaper
- water
- a ruler (mm)
- a marker

- 3 small planting containers
- sandy soil, clay, and loam
- pea, bean, or radish seeds
- a measuring cup
- a small garden shovel

What to do:

1. Spread the newspaper on the table. Then, using the masking tape and marker, label a planting container "sand", another "clay", and another "loam".

2. Fill each container 3/4 full with the soil on its label.

3. Place some seeds into each of the containers of soil. Add a little more soil on top of the seeds.

4. Add a little water to each container. Be sure it is the same amount for each container.

5. Place the containers in a warm, sunny place.

6. Make a prediction about which soil is going to be the best for growing plants. Record it on Worksheet 2.

7. Water your plants every other day, adding the same amount of water to each one. Make observations of the growing progress of your plants by measuring and illustrating them each time you water them.

8. Graph your observations and make a conclusion about the best soil for plant growth on Worksheet 5.

OTM2154 ISBN: 9781770789609
© On The Mark Press

Name: _____

Let's Predict

Which soil type is best for growing plants? Explain your thinking. _____

_____

_____

Let's Investigate

This is what I observed.

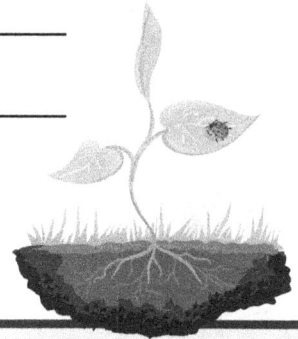

	Plant in Sandy Soil	Plant in Clay Soil	Plant in Loam Soil
Day 2	It is _____ mm tall.	It is _____ mm tall.	It is _____ mm tall.
Day 4	It is _____ mm tall.	It is _____ mm tall.	It is _____ mm tall.
Day 6	It is _____ mm tall.	It is _____ mm tall.	It is _____ mm tall.

OTM2154   ISBN: 9781770789609
© On The Mark Press

This is what I continued to observe.

	Plant in Sandy Soil	Plant in Clay Soil	Plant in Loam Soil
Day 8	It is _____ mm tall.	It is _____ mm tall.	It is _____ mm tall.
Day 10	It is _____ mm tall.	It is _____ mm tall.	It is _____ mm tall.
Day 12	It is _____ mm tall.	It is _____ mm tall.	It is _____ mm tall.
Day 14	It is _____ mm tall.	It is _____ mm tall.	It is _____ mm tall.

OTM2154   ISBN: 9781770789609
© On The Mark Press

Name:

This is what I continued to observe.

	Plant in Sandy Soil	Plant in Clay Soil	Plant in Loam Soil
Day 16	It is _____ mm tall.	It is _____ mm tall.	It is _____ mm tall.
Day 18	It is _____ mm tall.	It is _____ mm tall.	It is _____ mm tall.
Day 20	It is _____ mm tall.	It is _____ mm tall.	It is _____ mm tall.
Day 22	It is _____ mm tall.	It is _____ mm tall.	It is _____ mm tall.

OTM2154 ISBN: 9781770789609
© On The Mark Press

Name:

Using the data you collected, create a bar graph to show the plant growth in the three types of soil. Be sure to add a title and labels on your graph.

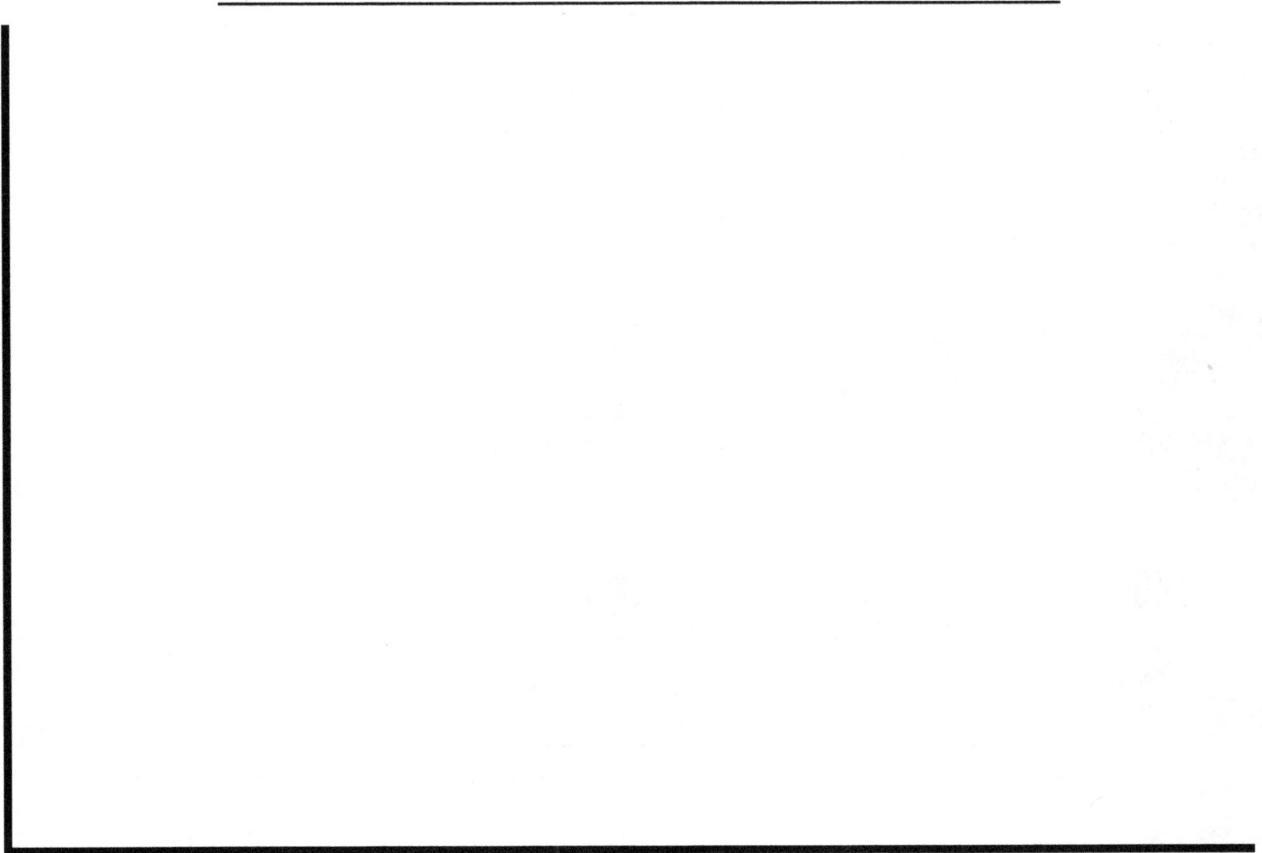

Let's Conclude

If you were planting a garden, what type of soil would you use? Use the data that you collected to support your answer. _____

_____

_____

_____

_____

OTM2154   ISBN: 9781770789609
© On The Mark Press

LIVING IN THE SOIL

LEARNING INTENTION:

Students will learn about how living things are interdependent with the soil they live in.

SUCCESS CRITERIA:

- locate and illustrate four livings things in or on soil in the environment
- make observations about the interdependence of soil and the things living in it
- discuss observations with a partner
- record observations in a chart, using words or diagrams to show interdependence
- choose one living thing in the soil and present findings to a group

MATERIALS NEEDED:

- a copy of "Alive in the Soil!" Worksheet 1, 2, 3, 4, and 5 for each student
- access to different soils in the neighborhood such as playground soil, garden/ flower bed soil, grassy field soil, lawn soil, beach/ riverbed soil, forest soil, or marsh soil
- small garden shovels (one per pair of students)
- magnifying glasses (one per pair of students)
- clipboards, pencils (one for each student)

PROCEDURE:

1. Explain to students that they will explore the living things that are in their neighborhood soils. Before going out:

   - discuss the meaning of 'interdependence'. It will be important for students to understand that living things in and on soil have a purpose for the soil, and the soil serves a purpose to the living things (animals and plants) in it.

   - it is important for students to understand that they must have a respect for all living things in the natural environment, this means that animals and plants are not to be harmed and that their habitats be maintained as much as possible

2. At this point, divide students into pairs. Give each pair a small garden shovel and a magnifying glass. Give each student a clipboard and pencil, and Worksheets 1, 2, 3, and 4. Read through the 'what to do' section with them to ensure their understanding of the task. Instruct students to draw detailed pictures of four living things that they find in or on soil. An option is to have students take photos of the living things, using an iPod, iPad, or camera. Once pairs have chosen a living thing, they are to examine its surroundings and discuss with their partner how it uses and depends on the soil, and how the soil benefits from the presence of the living thing. Notes on their discussion are to be recorded on Worksheets 3 and 4.

3. Return to the classroom. Give students Worksheet 5 to complete. Once it is completed, each student can share with the large group, or students can be divided into 2 or 3 smaller groups in order to share their ideas. If an iPad was used to take photos, they can be displayed using a data projector. The visual aid could lead to more class discussion.

DIFFERENTIATION:

Slower learners may benefit by omitting Worksheet 5. This expectation reduction will allow them more time to work on details for Worksheets 3 and 4.

For enrichment, faster learners could access the internet to research what they would like to know more about their living thing from their list on the bottom of Worksheet 5.

OTM2154 ISBN: 9781770789609
© On The Mark Press

Alive in the Soil!

You are going on another soil hunt! This time, you will look for things that are **living on and in the soils** in your neighborhood. You will need to have a close and careful eye to observe all the living things that depend on soil.

Here are some tips on where to look:

- in holes in the ground that show signs of animal life
- under rocks and logs
- between the grass and under fallen leaves
- underneath the soil

You'll need:

- a small garden shovel
- a clipboard and pencil
- a magnifying glass
- a partner
- a camera (optional)

What to do:

1. Use the small garden shovel and magnifying glass to help you look for signs of life in and on the soil.
2. Once you have found something, draw it on Worksheet 2, or take a picture of it.
3. When you find something, talk with your partner about:
   - how the living thing uses and depends on the soil
   - how the soil depends on the living thing that you found
4. Record notes from your talk on Worksheets 3 and 4.
5. Choose one living thing that you found in the soil that you would like to share with your classmates. On Worksheet 5, explain why you are choosing to share it.

OTM2154   ISBN: 9781770789609
© On The Mark Press

Name:

Draw and label four living things that you found in or on the soil in your neighborhood:

OTM2154   ISBN: 9781770789609
© On The Mark Press

Record the names of the living things you found. Next, describe how each uses and depends on the soil. Then, describe how the soil depends on each of those living things.

My partner and I found a _____

How does it use and depend on soil?

_____

_____

_____

How does the soil depend on this living thing?

_____

_____

_____

My partner and I found a _____

How does it use and depend on soil?

_____

_____

_____

How does the soil depend on this living thing?

_____

_____

OTM2154   ISBN: 9781770789609
© On The Mark Press

Name:

Continue recording the names of the living things you found. Describe how each uses and depends on the soil. Describe how the soil depends on each of those living things.

My partner and I found a ______________________

How does it use and depend on soil?

____

____

____

How does the soil depend on this living thing?

____

____

____

My partner and I found a ______________________

How does it use and depend on soil?

____

____

____

How does the soil depend on this living thing?

____

____

____

OTM2154 ISBN: 9781770789609

The living thing that I found in the soil that I would like to share with my classmates is:

_____

This living thing needs the soil to _____

_____

_____

_____

I find this living thing interesting because _____

_____

_____

_____

Some things that I would like **to know more about** this living thing are:

- _____

- _____

- _____

- _____

- _____

- _____

36

OTM2154 ISBN: 9781770789609
© On The Mark Press

AT WORK IN THE SOIL

LEARNING INTENTION:

Students will learn about the benefit of earthworms in making nutrient rich soil, and how composting provides nutrients to soil.

SUCCESS CRITERIA:

- describe where earthworms live, their purpose, and how they make nutrient rich soil
- create an earthworm farm
- make and record observations about earthworm activity
- make and record a conclusion about the benefit of having earthworms in gardens
- create a composter filled with organic matter
- make and record observations of the decomposition over time
- explain the uses and importance of compost
- make a connection to the environment by identifying nature's composting methods

MATERIALS NEEDED:

- ask each student to bring in a large wide-mouthed glass jar with a lid
- a copy of "Getting the Dirt on Earthworms!" Worksheet 1 for each student
- a copy of "A Home for the Earthworms" Worksheet 2 and 3 for each student
- a copy of "Classroom Composting" Worksheet 4, 5, and 6 for each student
- soil such as sand and loam or topsoil (enough to fill large jars for each student)
- a hammer and a nail, masking tape, a jug of water, a few small cups
- earthworms (2 or 3 per student)
- vegetable or fruit scraps
- large sheets of black construction paper (2 per student)
- pencils
- a large plastic bin with a lid
- garden soil
- vegetable and fruit scraps, egg shells, leaves, grass clippings, used coffee grinds or tea bags, or other organic material
- a garden shovel, a drill, a long stick for stirring
- access to water

PROCEDURE:

*This lesson can be done as one long lesson, or be divided into two shorter lessons. Composting should span over a two month period.

1. Using Worksheet 1 do a shared reading activity with the students. This will allow for reading practice and learning how to break down word parts in order to read the larger words in the text. Along with the content, discussion of certain vocabulary words would be of benefit for students to fully understand the passage.

   Some interesting vocabulary words to focus on are:

   - moist
   - nutrients
   - castings
   - burrows
   - digest
   - vegetation

2. Explain to students that they will create their own earthworm farms. Give them Worksheets 3 and 4, and the materials to create the farms. Read through the question, materials needed, and what to do sections on Worksheet 3 with the students to ensure their understanding of the task. Students will make and record observations of the earthworm farms as they are created and again 24 hours later. They will make a conclusion about the purpose earthworms have in creating nutrient rich soil.

3. Working as a large group, have students participate in creating and contributing to a classroom composter. Give students Worksheet 4. Read through the materials needed and what to do sections with them to ensure their understanding of the meaning of composting and the task they are required to participate in. Gather your materials and start building!

4. Give students Worksheets 5 and 6. They will illustrate the composting layers at the start of the project, and make observations of the decomposition of the organic matter over time on Worksheet 5. After a period of 8 weeks, students will illustrate the compost, and respond to questions on Worksheet 6. Emphasis should be made on the positive effects composting has on our soils and for our environment.

*An activity to enhance the learning about the importance of composting and necessity of decomposition, show students *The Magic School Bus* episode called "Meets the Rot Squad".

DIFFERENTIATION:

Slower learners may benefit by working as a small group with teacher direction to create their earthworm farms. It may also be beneficial for these learners to have another opportunity to re-read the information on Worksheet 1 in a small group. An additional accommodation is to work in a small group with teacher guidance to *orally respond* to the questions on Worksheet 6.

For enrichment, faster learners could create a graphic text about the adventures of their earthworms in the jar. An option is to use computer software to create it. A great computer program to assist with this is Comic Life. This is available on both PCs and Macs.

OTM2154   ISBN: 9781770789609
© On The Mark Press

Getting the Dirt on Earthworms!

Where Do They Live?

Earthworms make their homes in soil. Some earthworms live in the roots of grass under the lawn and others live in gardens among the vegetables. Some earthworms make their burrows under leaves among tree roots.

The earthworm's burrow is long, dark, and narrow. They stay in their burrows, where it is dark and moist, for most of the time. They come out at night to feed on leaves and grass, or other vegetation.

Worms help water flow through the soil!

What Do They Do?

The earthworm is like "nature's plow" because their burrowing of tunnels brings air to the roots of plants. They make the earth crumbly so that water can get into the soil easily.

When an earthworm eats and then digests its food, its castings add nutrients to the soil that plants need in order to grow.

Did You Know?

Farmers find earthworms very useful. They help to get air and water to the roots of their crops in the fields. Also, the earthworms' castings add nutrients to make the soil rich.

OTM2154   ISBN: 9781770789609
© On The Mark Press

Name:

A Home for the Earthworms

Some people make worm farms in order to make rich soil to add to their gardens at home. Let's give this a try!

You'll need:

- a large jar with a lid
- masking tape
- 2 or 3 earthworms
- a hammer and a nail
- soil (a mix of loam and sand)
- vegetable or fruit scraps
- a small cup of water
- 2 large sheets of black construction paper

What to do:

1. Fill the jar until it is three-quarters full of loam soil. Add a thin layer of sand to the top of the soil.

2. Throw some vegetable or fruit scraps on top of the sand.

3. Moisten the soil with a little bit of water. Then add the earthworms.

4. Using the hammer and a nail, **your teacher** will make some holes in the lid of the jar. Put the lid on the jar to close it.

5. On Worksheet 4, draw your observations of the jar and its contents.

6. Cover the sides of the jar with black construction paper so that no light can get in. **Worms like the dark!**

7. Leave the jar for a day.

8. The next day, remove the paper. Observe the changes in the jar. Record them on Worksheet 4.

OTM2154   ISBN: 9781770789609
© On The Mark Press

Let's Observe

Draw your observations of the jar and its contents.

This is what it looked like in the jar before it was covered up:	This is what it looked like in the jar after it was left for one day:

Let's Conclude

Explain the changes in the jar. What did the earthworms do?

_____

_____

_____

_____

Classroom Composting

Composting is a process that breaks down things like food waste, leaves, grass clippings, and wood bits into humus. **Compost** is rich in nutrients. When it is put into gardens, it helps to grow healthy new plants.

You'll need:

- a large plastic bin with a lid
- leaves, grass clippings
- vegetable and fruit scraps
- a long stick for stirring
- garden soil
- a garden shovel
- a drill
- access to water
- egg shells

What to do:

1. Put some soil in the bottom of the plastic bin.
2. Add vegetable or fruit scraps, egg shells, leaves, and grass clippings. Moisten the soil with a bit of water.
3. Repeat steps 1 and 2 to make layers. Illustrate it on Worksheet 5.
4. Using the drill, **your teacher** will make some holes in the lid of the bin. Put the lid on the bin to close it, and place it outside.
5. Use the long stick to stir your compost pile each week. Add organic materials when available. Make and record your observations of the changes you notice on Worksheet 5.
6. After several weeks, you will have created some humus that you can add to a garden to grow new plants. Illustrate it on Worksheet 6.

OTM2154   ISBN: 9781770789609
© On The Mark Press

Name:

Draw a diagram of the composter. Show the layers of ingredients you put in it.

Let's Observe!

As organic matter "cooks", explain the changes that you are noticing.

_____

_____

_____

_____

_____

_____

Name:

Illustrate the humus you have created.

How will you use it?

_____

_____

_____

_____

_____

_____

_____

_____

Make a list of reasons why people should compost.

• _____

• _____

• _____

• _____

Challenge question:

How does composting happen in nature?

_____

_____

_____

_____

OTM2154   ISBN: 9781770789609
© On The Mark Press

EROSION

LEARNING INTENTION:
Students will learn about the effects of water on soil.

SUCCESS CRITERIA:
- make predictions about the effect of water on soil
- observe and describe the effect that water has on different types of soil
- gather data and record observations in a chart
- make conclusions about soil erosion using graphs and written descriptions
- make connections about the effects of water on soils in the environment

MATERIALS NEEDED:
- a copy of "Experimenting with Soil Absorption" Worksheet 1, 2, and 3 for each student
- a copy of "Signs of Erosion" Worksheet 4 for each student
- a copy of "Controlling Erosion" Worksheet 5, 6, and 7 for each student
- samples of soils such as sand, clay, silt, loam, and topsoil (enough for 4 or 5 groups)
- a pan of water, 3 cotton cloths or coffee filters, 3 cups, 3 graduated cylinders, a small garden spade or large spoon (for each group of students)
- 2 large plastic bins (deep and rectangular), soil to fill the bins, a piece of grass sod, an empty milk container (cardboard), an empty jug, 4 litres of water (for each pair of students)
- a sharp knife or scissors
- pencils, clipboards
- chart paper, markers

PROCEDURE:
*This lesson can be done as one long lesson, or be divided into three shorter lessons.

1. Explain to students that they will do an experiment to see what effect water has on different types of soils. Give them Worksheet 1. Read through the question, materials needed, and what to do sections with the students to ensure their understanding. Divide students into pairs or small groups. Give them Worksheets 2 and 3, and the materials to conduct the experiment. Students will make and record observations of the effects water has on each of the soils, then make conclusions and a connection. Follow-up discussion would be beneficial to ensure students understand the concept of soil erosion by water. Discussion could include examples of soil erosion or flooding in places in the world.

2. Give students Worksheet 4, a clipboard, and a pencil. Take them out on a walk through the neighborhood to look for signs of soil erosion. (Some areas to look for may be along road shoulders, along a stream bank, under an eaves trough spout or sprinkler.) Upon completion of Worksheet 4, have students share some of their findings with the large group. Engage students in a discussion about the dangerous/ negative effects that moving water has on our landscape.

3. Explain to students that they will do an experiment to see what effect plant life has on soil erosion. Give them Worksheet 5. Read through the question, materials needed, and what to do sections with the students to ensure their understanding of the task. (To save time, have milk containers available that already have holes in the bottom of them.) Divide students into pairs or small groups. Give them Worksheets 6 and 7, and the materials to conduct the experiment. Students will make and record observations of the effect grass has on preventing soil erosion. **(Grasses have fibrous roots that spread out into the soil in**

many directions. This helps to hold the soil together, so water does not easily erode it.) Students will make a conclusion about and then work in pairs to make connections between human activity/efforts and soil erosion.

DIFFERENTIATION:

Slower learners may benefit by working as a small group with teacher direction and support in order to provide accurate measurements while experimenting with the soils. This would result in one set of data to graph, which could be done together, using chart paper and markers. A further accommodation may be to change the 'making connections' section on Worksheet 3 to a discussion item within the small group if time permits.

For enrichment, faster learners could work with a partner or in a small group to discuss which soils would be most appropriate to use for making pottery, building earth shelters or sand castles, road construction, landscaping, etc. Then give reasons for their choices.

OTM2154   ISBN: 9781770789609
© On The Mark Press

Experimenting with Soil Absorption

Question: What effect does water have on different types of soils?

You'll need:

- a pan of water
- 3 graduated cylinders
- a small garden spade
- 3 cups
- 3 cotton cloths or coffee filters
- soils such as sand, clay, and loam

What to do:

1. Make a prediction to the question and record it on Worksheet 2.

2. Using the gardening spade, scoop some sand and put it on a cotton cloth. Wrap up the cloth. Repeat this step using the clay and then using the loam.

3. Take the cloth wrapped with sand in it, and carefully dip it into the pan of water.

4. Pull the cloth out of the water and let it drip over a cup. Make observations about the amount of water you see dripping from the cloth filled with sand. Draw what you saw in the chart on Worksheet 2.

5. Repeat steps 3 and 4 using the cloth filled with clay, and the cloth filled with loam.

6. Pour the water from each cup into the graduated cylinders. Record the amount of water in each cylinder in the chart on Worksheet 2.

7. Make conclusions and connections about the effect of water on each soil type. Record them on Worksheet 3.

OTM2154   ISBN: 9781770789609
© On The Mark Press

Let's Predict

What effect does water have on different types of soils?

_____

_____

_____

Let's Observe

Cloth filled with sand	Cloth filled with clay	Cloth filled with loam
This is what it looked like when I pulled the cloth filled with **sand** out of the water:	This is what it looked like when I pulled the cloth filled with **clay** out of the water:	This is what it looked like when I pulled the cloth filled with **loam** out of the water:
When I poured the water from the cup into the graduated cylinder, it measured: _____	When I poured the water from the cup into the graduated cylinder, it measured: _____	When I poured the water from the cup into the graduated cylinder, it measured: _____

OTM2154   ISBN: 9781770789609
© On The Mark Press

Name:

Using the data you collected, create a bar graph to show the water absorption of each soil. Be sure to add a title and labels on your graph.

Let's Connect It!

If you wanted to stop a river from overflowing, you could place bags filled with soil along the riverbanks. What kind of soil would you use to fill the bags? Explain your thinking.

_____

_____

_____

_____

_____

Signs of Erosion

Take a walk in your neighborhood. Look for signs of soil erosion. Illustrate four signs of soil erosion. Describe how the erosion happened.

OTM2154   ISBN: 9781770789609
© On The Mark Press

Controlling Erosion

Question: Do grass and plants help to prevent erosion?

You'll need:

- 2 large plastic bins (deep and rectangular)
- an empty milk container (cardboard)
- a piece of grass sod

- a sharp knife or scissors
- 4 litres of water
- a partner
- a garden shovel
- an empty jug
- soil

What to do:

1. Make a prediction and record it on Worksheet 6.

2. Using the garden shovel, fill both of the large plastic bins with some soil. Shape it to form one side of a hill.

3. In one of the bins, place the piece of grass sod onto the soil mound. This will represent a grassy hillside.

4. **Your teacher** will make some holes in the bottom of the milk container.

5. Your partner will hold the milk container over top of the bin of only soil.

6. You will use the jug to pour 2 litres of water in through the milk container, while your partner moves it over top of the hill of soil. This will represent rain.

7. Repeat steps 5 and 6, using the bin of soil with grass sod.

8. Observe the amount of water that collects at the bottom of each hill, and the consistency and color of the water. Record your observations.

9. Make conclusions and connections about soil erosion on Worksheet 7.

Let's Predict

Do grass and plants help to prevent erosion? Explain your thinking. _____

_____

_____

Let's Observe

Bin with **soil** hillside	Bin with **grassy** hillside
Illustration of the **soil** hillside after it rained:	Illustration of the **grassy** hillside after it rained:
Describe the water collection at the bottom of the hill after it had rained. _____ _____ _____ _____ _____ _____	Describe the water collection at the bottom of the hill after it had rained. _____ _____ _____ _____ _____ _____

OTM2154   ISBN: 9781770789609

Name:

Let's Conclude

Do grass and plants help to prevent erosion?
Explain your thinking. _____

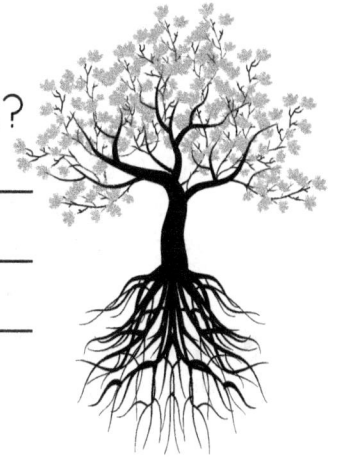

_____

_____

Let's Connect It!

Think | Pair | Share

With a partner, do some thinking and sharing of ideas about things that cause soil erosion.

Complete the web below by adding things that humans can do **to prevent** our landscape from eroding.

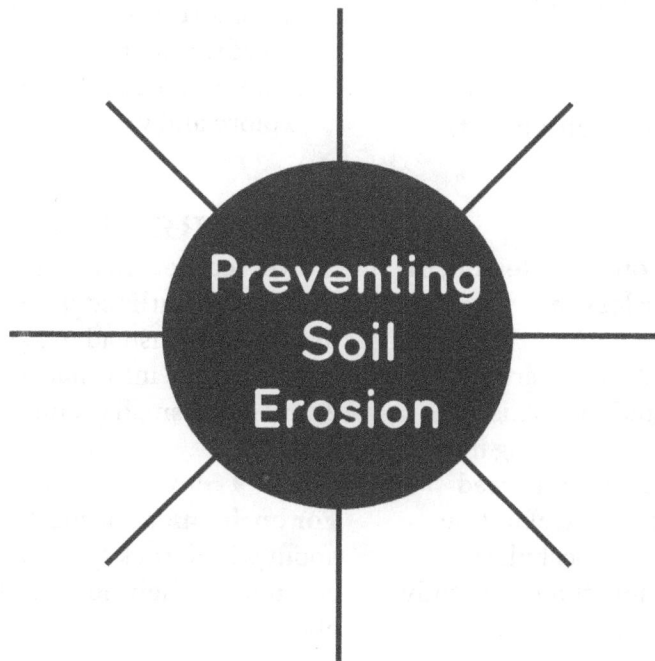

Preventing
Soil
Erosion

ROCK TYPES

LEARNING INTENTION:
Students will learn about different rock types and discover the types that are where they live.

SUCCESS CRITERIA:
- describe the different types of rocks
- locate and illustrate four different rock types in the neighborhood
- identify rocks as either igneous, sedimentary, or metamorphic
- choose and present findings about one type of rock to a group

MATERIALS NEEDED:
- a copy of "Types of Rocks" Worksheet 1, 2, and 3 for each student
- a copy of "What's Rocking Your Neighborhood?" Worksheet 4 for each student
- access to different areas in the neighborhood in order to collect different types of rocks
- plastic bags or small cardboard boxes (one per student)
- magnifying glasses (one per pair of students)
- clipboards, pencils (one for each student)

PROCEDURE:
*This lesson can be done as one long lesson, or be divided into two shorter lessons.

1. Using Worksheet 1, 2, and 3, do a shared reading activity with the students. This will allow for reading practice and learning how to break down word parts in order to read the larger words in the text. Along with the content, discussion of certain vocabulary words would be of benefit for students to fully understand the passage.

   Some interesting vocabulary words are:

   - igneous
   - sedimentary
   - metamorphic
   - pressure
   - magma
   - weathering
   - compact
   - volcanic eruption

2. Explain to students that they will go on an exploration to discover some rock types that are in their neighborhood. Give each student a plastic bag or small cardboard box. Instruct them to collect about four interesting rocks that they would like to bring back to the classroom. (These rocks will become a classroom rock collection, to be used throughout the unit.)

3. Return to the classroom. Give students Worksheet 4 to complete. Options for this activity:
   - students can work individually using the rocks that they collected
   - students can work in a small group, pooling their rocks together, then choosing four rocks to illustrate on Worksheet 4

4. Students can choose one rock that they would like to share with the large group. They should tell what type of rock it is, and explain why they chose to share it. Collect all rocks at the end of this activity and display them on a counter/table in the classroom for students to explore and work with in later activities.

DIFFERENTIATION:
Slower learners may benefit by not participating in step 4 as outlined in the procedure section, but use the time instead to have another opportunity to re-read the information on Worksheets 1, 2, and 3 in a small group with teacher or peer support.

For enrichment, faster learners could write about which rock type they think is most common in their neighborhood, explaining their choice.

OTM2154   ISBN: 9781770789609
© On The Mark Press

Types of Rocks

There are three different types of rocks. These types are igneous rocks, sedimentary rocks, and metamorphic rocks. Let's read more about them!

Igneous

Igneous rocks have been formed by heat. This type of rock forms when molten rock, called magma, rises to the earth's surface through cracks caused by earthquakes and volcanic eruptions. When magma reaches the earth's surface, it becomes hot lava. This lava will cool into cold hard rock, which is known as molten rock. Igneous rocks are formed.

Basalt

Granite

Usually these rocks are heavy, but there are some that have air bubbles, which makes them very lightweight.

Sedimentary

Rocks are broken up by a natural process called "weathering". Weathering happens when rocks are washed away by water or blown away by wind, causing them to be broken into smaller pieces. These smaller pieces of rock settle into layers on land, in riverbeds, lakes, and oceans.

The weight of the layers of sand and water presses down on the bottom layer and squeezes out the water, which compacts it. The grains of sand and tiny rocks begin to stick together, eventually becoming stone. This is **sedimentary** rock.

Sandstone, limestone, and shale are the common types of sedimentary rock.

Sandstone

Limestone

Shale

OTM2154   ISBN: 9781770789609
© On The Mark Press

Name:

Metamorphic

Metamorphic means a change in form. Metamorphic rocks were all once igneous or sedimentary rocks. Hot masses of magma sometimes slowly push out of the earth's crust. The heat and pressure of the magma invades an igneous or sedimentary rock, causing it to change its form.

When limestone, a soft sedimentary rock is invaded by a mass of magma, it is changed into marble, a very hard stone.

Limestone

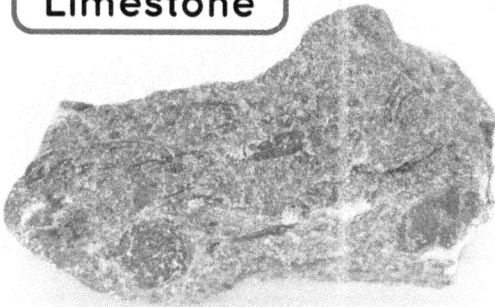

Recrystallization
(caused by heat)

Marble

Did you know?

The Acasta Gneiss is a large metamorphic rock outcrop in the Northwest Territories, Canada. This rock outcrop is located about 300 kilometres north of Yellowknife. It is believed to be the oldest known rock in the world, about 4 billion years old!

OTM2154 ISBN: 9781770789609
© On The Mark Press

What's Rocking Your Neighborhood?

In the boxes below, illustrate some rocks that you find.
Label them as **igneous**, sedimentary, or **metamorphic**.

OTM2154   ISBN: 9781770789609
© On The Mark Press

EXPLORING MINERALS

LEARNING INTENTION:
Students will learn about minerals and their properties.

SUCCESS CRITERIA:
- describe different properties of minerals
- recognize the difference between a rock and a mineral
- conduct a study to describe a rock according to its properties
- share findings about one rock with a classmate

MATERIALS NEEDED:
- a copy of "Minerals" Worksheet 1 and 2 for each student
- a copy of "Rock vs. Mineral" Worksheet 3 for each student
- a copy of "A Rock Study" Worksheet 4 for each student
- a collection of different types of minerals (this can be purchased from an educational store/ website, or may be borrowed from a school or central library)
- 6 measuring tapes, a weight scale, 2 balance scales, copper coins, nails, pieces of glass, 6 steel files or knives
- magnifying glasses (one per student)
- pencils, chart paper, markers

PROCEDURE:
*This lesson can be done as one long lesson, or be divided into two shorter lessons.

1. Using Worksheet 1and 2, do a shared reading activity with the students. This will allow for reading practice and learning how to break down word parts in order to read the larger words in the text. Along with the content, discussion of certain vocabulary words would be of benefit for students to fully understand the passage.

Some interesting vocabulary words are:

- luster
- cleavage (to split)
- nonmetallic
- metallic
- substance
- surface

2. Set up mineral exploration centres in the classroom so that groups of students can rotate and use magnifying glasses to get a close look at some mineral samples. Class discussion could follow to allow students to share their findings. Encourage students to comment on luster, color, shape, hardness, etc.

3. Give students Worksheet 3. Read through information as a large group. "Can you guess" section should be done orally. Answers to Worksheet 3:

   (top row left to right) – mineral (fluorite), rock (gneiss), mineral (quartz), mineral (pyrite)
   (bottom row left to right) – rock (sandstone), mineral (amazonite), rock (marble)

4. Give students Worksheet 4. They will choose a rock from the classroom rock collection to study. Once their study is complete, they can share their findings with a friend.

DIFFERENTIATION:
Slower learners may benefit by working as a small group with teacher direction and support in order to provide accurate observations while conducting the rock study. This would result in one record of information, which could be done together, using chart paper and markers. This could be shared with the large group afterwards, if time permits.

For enrichment, faster learners could choose another rock to study.

OTM2154 ISBN: 9781770789609
© On The Mark Press

Minerals

All rocks on Earth contain minerals. Minerals occur in nature and they have a crystal structure and contain chemicals. Minerals have certain characteristics, such as luster, color, cleavage, and hardness. Let's read more about them!

The **luster** of a rock means it may be shiny or dull. The minerals in the rock may be metallic or nonmetallic. Minerals that have a metallic luster make the rock shiny.

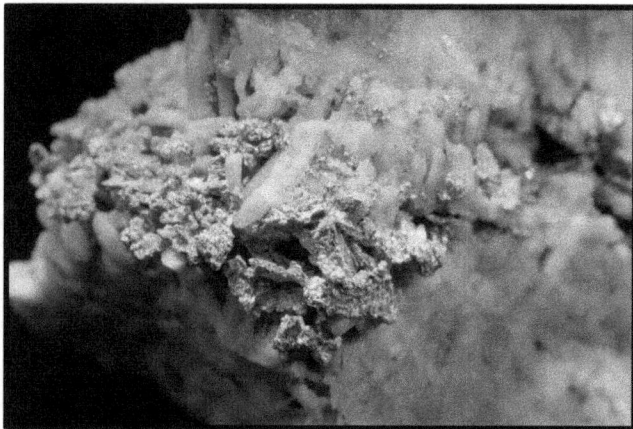

Gold is a metallic mineral. It has a shiny surface. Can you spot the gold in this rock?

Talc is a nonmetallic mineral. It has a pearly surface.

The **color** of a mineral depends on the materials that make up the crystals. For example, pure quartz has colorless crystals, but small amounts of other substances can give it a pink or green tint. Amethyst is a purple variety of quartz. Its purple color is caused by the substance iron. The amount of iron in it determines how bright in color it can be.

OTM2154   ISBN: 9781770789609
© On The Mark Press

Cleavage describes how the mineral breaks. When a mineral is cleaved, it may break with one, two, three, or more surfaces. It will break along smooth planes that are close to weak bonding areas within the crystal.

Mica

Mica breaks down in one direction. When it is split, it forms thin sheets.

Halite

Halite, which is a salt crystal, has cubic cleavage. When it is split, it looks like a cube.

Some minerals like **quartz** do not have cleavage and will only break into uneven pieces with rough surfaces.

Quartz

The hardness of minerals can be tested by scratching one mineral with another. In 1812, Friedrich Mohs created a scale of hardness. He took ten minerals and numbered them from one, the softest, to ten, the hardest. No mineral on the scale can be scratched by any one that is softer, but it can scratch all that are softer.

Moh's Scale of Hardness

1. **Talc**	(scratch with fingernail)	6. **Feldspar**	(scratch with a knife)
2. **Gypsum**	(scratch with fingernail)	7. **Quartz**	(scratches glass)
3. **Calcite**	(scratch with copper coin)	8. **Topaz**	(scratches glass)
4. **Fluorite**	(scratch with a steel nail)	9. **Corundum**	(scratches topaz)
5. **Apatite**	(scratch with a steel nail)	10. **Diamond**	(scratches another diamond)

OTM2154 ISBN: 9781770789609
© On The Mark Press

Name:

Rock vs. Mineral

Are you wondering what the difference is between rocks and minerals? Simply put, **a mineral is made of the same substance throughout. A rock is made up of more than one mineral.**

You have read about and examined some minerals such as feldspar, quartz, and mica. As you looked closely at each, you saw that they were of one substance. Mixed together in different amounts with different types of rocks, they can look very different from their original mineral form. Gneiss, granite, and sandstone are rocks that all have bits of feldspar, quartz, and mica in them.

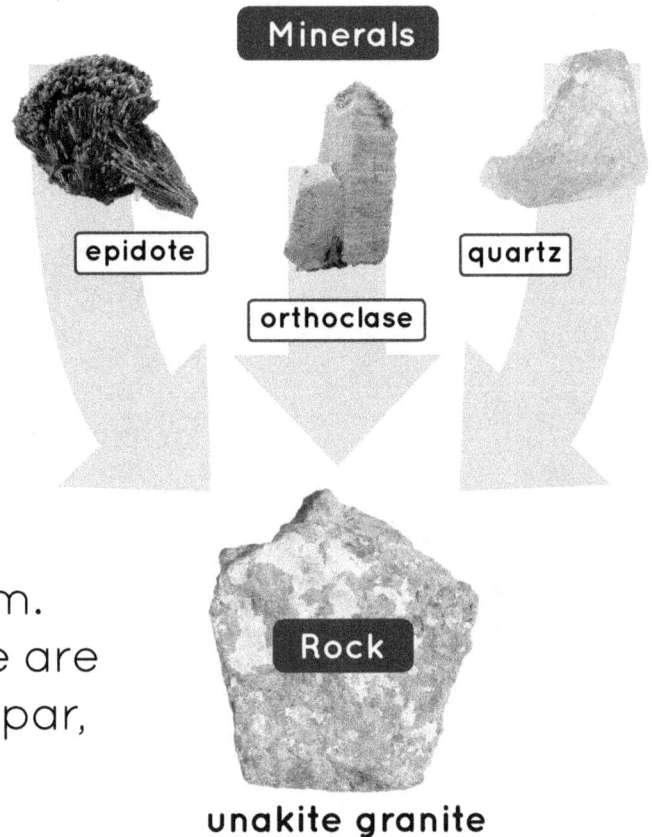

Minerals

epidote

orthoclase

quartz

Rock

unakite granite

Can You Guess?

Is it a rock or a mineral?

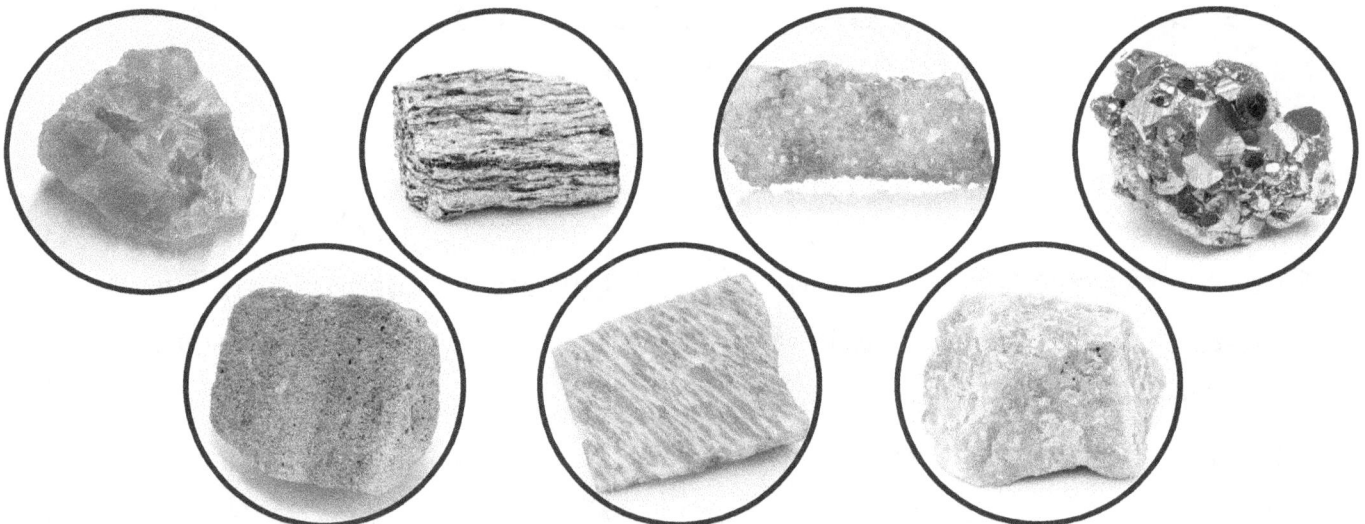

OTM2154   ISBN: 9781770789609

A Rock Study

You have learned a lot about rocks and the minerals in them. It's time to do a rock study!

Choose a rock from your classroom rock collection. Examine its size, weight, shape, color, texture, and hardness. Use these clues to help you determine what type of rock it is and what minerals it may contain.

Characteristics	Qualities about the Rock
Size: small, medium, large? measurement around?	
Weight: put it on the scale, what does it weigh?	
Shape: round, jagged, pointed?	
Color: what colors does it have?	
Luster: shiny, dull, metallic, nonmetallic?	
Texture: rough, bumpy, smooth?	
Hardness: can it be scratched? what scratches it? rate on Moh's Scale?	
Type of rock: igneous? sedimentary? metamorphic?	
Minerals: what minerals are in it?	

FUN WITH ROCKS

LEARNING INTENTION:
Students will learn about the presence of carbonates, and how rocks become eroded.

SUCCESS CRITERIA:
- conduct a test to determine the presence of calcite in a sedimentary rock
- record observations in a chart using pictures
- make a conclusion about the type of sedimentary rock
- replicate the effects of moving water over rocks
- record observations in a chart using pictures and words
- make a conclusion about the changes in rock formation
- make a connection to the environment

MATERIALS NEEDED:
- a copy of "Testing for Limestone" Worksheet 1 and 2 for each student
- a copy of "Shake Down to Break Down!" Worksheet 3 and 4 for each student
- 3 different sedimentary rocks, 3 glass jars, 300 ml of white vinegar, 2 magnifying glasses, a measuring cup (for each pair of students)
- 9 sedimentary rocks, 9 igneous or metamorphic rocks, 3 empty coffee cans with lids, 3 clear plastic containers, a large spoon with holes, a sieve, a jug of water (for each group)
- masking tape, markers, chart paper, pencils

PROCEDURE:
*This lesson can be done as one long lesson, or be divided into two shorter lessons.

1. Give students Worksheets 1 and 2 and materials to conduct the experiment (ensure that each pair of students has a piece of limestone). Read through the materials needed and what to do section to ensure their understanding of the task. Upon completion of the experiment, students will determine if any of their sedimentary rocks are limestone.

Some interesting facts about why scientists would want to know if sedimentary rock is limestone and contains calcite:

- it is used in the construction industry
- calcite is used in antacid tablets to reduce stomach acid
- calcite is used as a whitening agent in paint, and to remove stains in clothing
- ground limestone is sprayed on the walls in coal mines to reduce the dust in the air, it also reflects light in the dark mine

2. Give students Worksheets 3 and 4 and materials to conduct the experiment. Read through the materials needed and what to do section to ensure their understanding of the task. Upon completion of the experiment, students will determine how rocks change form due to erosion caused by moving water.

DIFFERENTIATION:
Slower learners may benefit by working as a small group with teacher direction and support in order to provide accurate observations while conducting the experiments. This would result in one record of information, which could be done together, using chart paper and markers. An additional accommodation may be to only have them conduct one of the two experiments.

For enrichment, faster learners could describe some common uses of rocks and minerals, explaining how they are used within the school, at home, or in the community.

OTM2154   ISBN: 9781770789609
© On The Mark Press

Testing for Limestone

You have learned that limestone is a sedimentary rock. Can you tell it apart from other sedimentary rocks? Let's try this simple test to see if a sedimentary rock is limestone!

You'll need:

- 3 different looking sedimentary rocks

- 3 glass jars

- 300 ml of white vinegar

- a magnifying glass

- a measuring cup

What to do:

1. Place a sedimentary rock in each of the glasses.

2. Pour 100 ml of vinegar into the measuring cup.

3. Pour the vinegar over the sedimentary rock in the first glass.

4. Using your magnifying glass, observe what happens in the glass.

5. Record your observations on Worksheet 2.

6. Repeat steps 2 through 5 for each of the remaining rocks.

7. Make a conclusion about which of the rocks may be limestone. Record it on Worksheet 2.

Name:

Let's Observe

Rock #1	Rock #2	Rock #3
This is what it looked like when I poured the vinegar over it:	This is what it looked like when I poured the vinegar over it:	This is what it looked like when I poured the vinegar over it:
Did you see any tiny bubbles? _____	Did you see any tiny bubbles? _____	Did you see any tiny bubbles? _____

Bubbling is a sign of a chemical reaction. Vinegar and the mineral **calcite** will create carbon dioxide. If you saw bubbles then the rock is limestone because it contains the mineral calcite.

I ♥ bubble baths!

Let's Conclude

Were any of the rocks you tested limestone? Explain your results. _____

_____

_____

OTM2154 ISBN: 9781770789609
© On The Mark Press

Shake Down to Break Down!

You have learned that weathering of rocks happens when they are washed away by water or blown away by wind, causing them to be broken into smaller pieces. This is **erosion**. Let's experiment with this idea!

You'll need:

- 9 sedimentary rocks
- 9 igneous or metamorphic rocks
- 3 empty coffee cans with lids
- 3 clear plastic containers
- a large spoon with holes, and a seive
- a marker
- a jug of water
- masking tape
- a partner

What to do:

1. Put 3 sedimentary rocks into each of the coffee cans. Then put 3 igneous or metamorphic rocks into each of the coffee cans.

2. Pour some water into each of the coffee cans so that it covers the rocks. Put the lid on each can.

3. Using the masking tape and marker, create a label for the first can that says "no shakes", a label for the second can that says "50" shakes, and a label for the third can that says "500 shakes". Do this same step for each of the clear plastic containers.

4. Shake the can labeled "50 shakes", 50 times. Shake the can labeled "500" shakes, 500 times. (Take turns).

5. Remove the rocks from each container, keeping the piles separate.

6. Pour the water from each can through the sieve, into its matching labeled plastic container. Put the remaining rocks in the sieve in its matching pile.

7. Record your observations and conclusions about the rocks and water on Worksheet 4.

Let's Observe

"No Shakes"	"50 Shakes"	"100 Shakes"
This is what **the rocks** looked like when I poured the water through the sieve:	This is what **the rocks** looked like when I poured the water through the sieve:	This is what **the rocks** looked like when I poured the water through the sieve:
Describe what **the water** looked like once it had passed through the sieve. _____ _____ _____ _____	Describe what **the water** looked like once it had passed through the sieve. _____ _____ _____ _____	Describe what **the water** looked like once it had passed through the sieve. _____ _____ _____ _____

Let's Conclude

What caused the rocks to change form?

_____

_____

Use your results to explain what happens to rocks that are carried down the river by water current.

_____

_____

OTM2154   ISBN: 9781770789609
© On The Mark Press

OUR SOLAR SYSTEM

LEARNING INTENTION:

Students will learn about the unique features of each planet in our solar system.

SUCCESS CRITERIA:

- identify and describe unique features of the eight planets in our solar system
- participate in a quiz of knowledge about our solar system
- create a model of the solar system

MATERIALS NEEDED:

- a copy of "Our Solar System" Worksheet 1 for each student
- a copy of "The Planets" Worksheet 2, 3, 4, and 5 for each student
- a copy of "Planet Trivia" Worksheet 6, 7, and 8 for each pair of students
- a copy of "A Model of the Solar System" Worksheet 9 for each student
- 9 Styrofoam balls (4 large, 4 small, 1 extra large), a piece of flat Styrofoam (15cm x15cm), a large Styrofoam cup, 13 bamboo skewers (for each pair of students)
- assorted colors of paint and paint brushes, string, rulers, masking tape, scissors, markers, pencils

PROCEDURE:

***This lesson can be done as one long lesson, or be divided into three shorter lessons.**

1. Using Worksheet 1, 2, 3, 4, and 5, do a shared reading activity with the students. This will allow for reading practice and learning how to break down word parts in order to read the larger words in the text. Along with the content, discussion of certain vocabulary words would be of benefit for students to fully understand the passage.

Some interesting vocabulary words to focus on are:

- terrestrial
- dwarf
- surrounded
- erupt
- particles
- astronomy
- diameter
- surface
- rotate
- span
- atmosphere
- canyon
- horizontal
- hurricane
- orbit

2. Divide students into pairs. Give them Worksheets 6, 7, and 8. They will cut out the trivia cards, creating two more questions of their own to add to the game. Each student will take a turn asking their partner a planet trivia question (correct answer is highlighted).

3. Give pairs of students Worksheet 9 and the materials to create a model of the solar system. Read through the materials needed and what to do sections to ensure their understanding of the task before they begin. This project may span over a couple of days, as the painted planets will need time to dry before assembly can begin.

DIFFERENTIATION:

Slower learners may benefit by working as a small group with teacher support in order to participate in the planet trivia game. They may need to reference the information as they provide answers to the questions. This would allow these learners to have another opportunity to re-read the information on Worksheets 1, 2, 3, 4, and 5 in a small group.

For enrichment, faster learners could choose a planet and create a travel brochure to convince people to visit it. (Provide sample brochures so students understand the purpose and layout). Students' travel brochures should include:

location, size, composition, appearance (unique features), temperature, weather, moons, length of day and year, how long it takes to travel there from Earth, its distance from the sun

OTM2154   ISBN: 9781770789609
© On The Mark Press

Our Solar System

Have you ever wondered what is out there, beyond Earth?

Earth is just one of **eight planets** and **five dwarf planets** that orbit the Sun, along with many other objects such as asteroids, comets, meteoroids, and satellites.

The four planets closest to the sun are Mercury, Venus, Earth, and Mars. They are called the **terrestrial planets** because they have solid rocky surfaces.

The four large planets Jupiter, Saturn, Uranus, and Neptune are called the **gas giants** because they have a thick atmosphere of hydrogen and helium.

Did You Know?

Pluto was the ninth planet in our solar system until 2006 when it was demoted to a **dwarf planet**. A professor discovered an icy object about the same size as Pluto, out beyond its orbit. This object is the dwarf planet called Eris. It turns out that Pluto is just a fraction of the icy objects in its realm, so it is now considered a dwarf planet.

OTM2154   ISBN: 9781770789609
© On The Mark Press

The Planets

Mercury is the closest planet to the sun. Its diameter is 4880 kilometres. It can get as hot as 430 °C or as cold as -170 °C. It has no weather patterns or water. On its surface there are mountains and craters, and it has a heavy iron core.

It takes Mercury 88 days to go around the sun once. This means that one year on Mercury is equal to 88 Earth days.

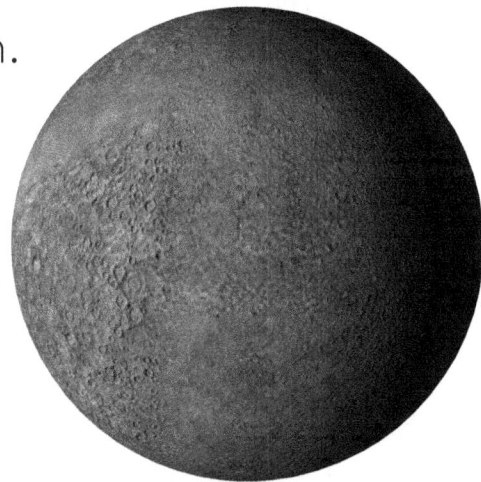

Venus is the second planet from the sun. It is 12 104 kilometres in diameter. It is surrounded by clouds containing a poisonous gas called sulfuric acid. These clouds trap the sun's heat. As a result, the temperature on Venus is about 450 °C. This hot temperature causes it to glow red.

It takes Venus 225 days to orbit the sun. This means one year on Venus is equal to 225 Earth days.

The surface of Venus is rocky and desert-like, hot and dry. There is no water. Much of its surface is covered by hardened lava flows from volcanoes that erupted a long time ago.

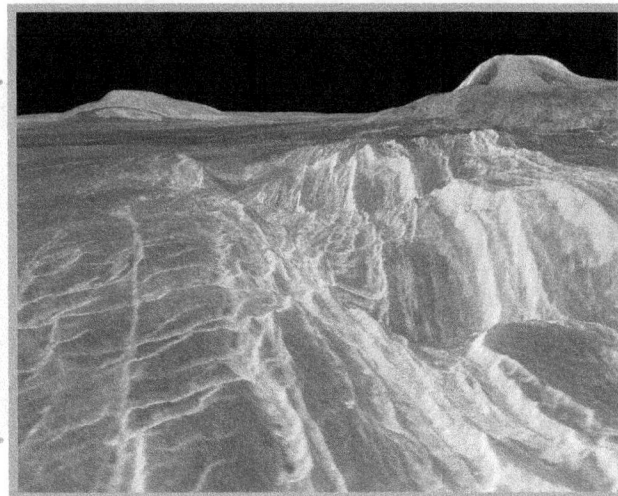

OTM2154 ISBN: 9781770789609
© On The Mark Press

Name:

The **Earth** is the third planet from the sun. Its diameter is 12 756 kilometres. The Earth has hot deserts and cold, icy poles. The average temperature on Earth is 22 °C, which makes it ideal for life forms. It is the only plant known to have life on it. The Earth's atmosphere contains mainly oxygen and carbon dioxide, and it protects life forms from the rays of the sun.

Mars is the fourth planet from the sun. It is 6 787 kilometres in diameter. It has two moons, Phobos and Deimos. Its temperature ranges from -120 °C during the winter and 25 °C in the summer. It has a thin atmosphere made mostly of carbon dioxide.

Mars is nicknamed the **Red Planet** because of the red color caused by rusted iron in its dusty, rocky surface. Mars has big volcanoes, and many channels and canyons on its surface which could have been eroded by water a long time ago. Presently there is no liquid water on Mars, only frozen water at its polar ice caps.

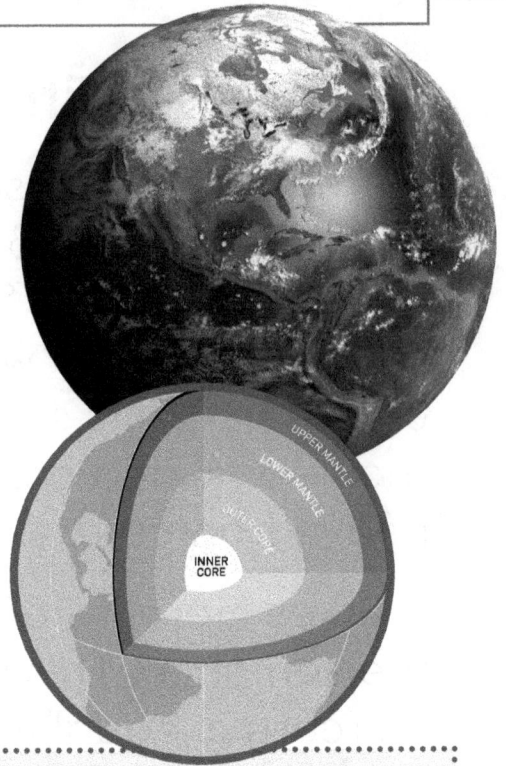

UPPER MANTLE

LOWER MANTLE

OUTER CORE

INNER CORE

Almost 3/4 of the Earth is covered by water. The rest of Earth is land. The land and water are above the Earth's crust.

Under the Earth's crust is a deep layer of rock called the mantle. Under this is the core, which is made up of mostly iron and nickel.

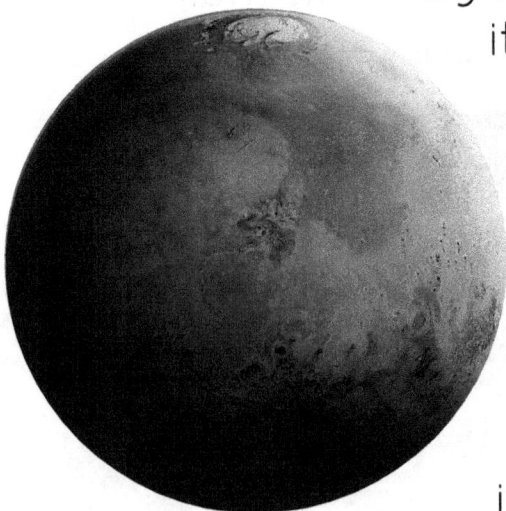

OTM2154 ISBN: 9781770789609
© On The Mark Press

Name:

Jupiter is the fifth planet from the sun. The diameter of this huge planet is 142 200 kilometres. Jupiter is a cold planet, with the average temperature being -150 °C.

Its atmosphere and surface are mostly made up of hydrogen and helium. Jupiter has a great red spot, which is a giant mass of swirling gases. This spot is like a hurricane storm that has been raging for hundreds of years.

Jupiter has 50 known moons circling around it. It also has three rings, made of fine particles like dust, that are more easily seen when backlit by the sun.

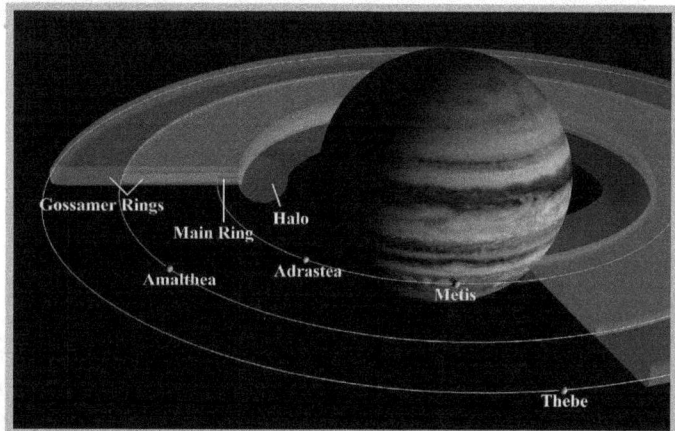

Gossamer Rings
Main Ring
Halo
Amalthea
Adrastea
Metis
Thebe

Saturn is the sixth planet from the sun. This large ringed planet has a diameter of 119 300 kilometres. It is a cold planet, with the average temperature being -180 °C.

Its atmosphere and surface are mostly made up of the gases hydrogen and helium. Saturn's ring system is actually made of icy objects flying around the planet. This ring system spans out hundreds of thousands of kilometers from the planet. Saturn has 53 known moons within its magnetic field. It takes Saturn 29.5 years to orbit the sun one time.

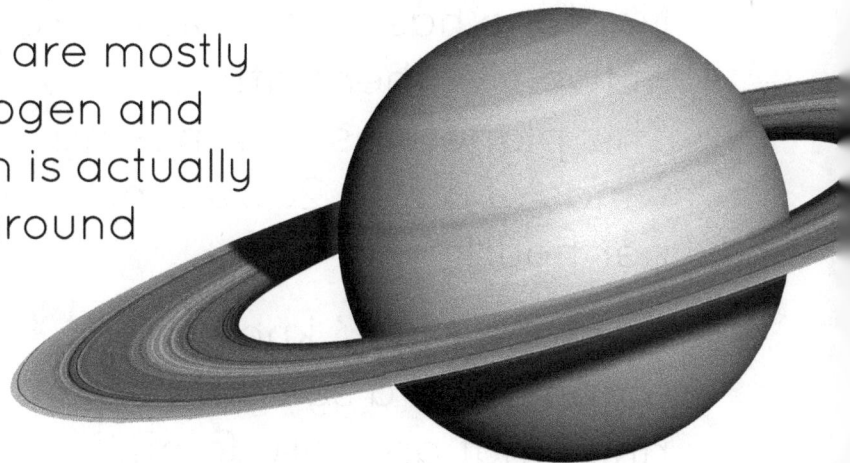

OTM2154   ISBN: 9781770789609
© On The Mark Press

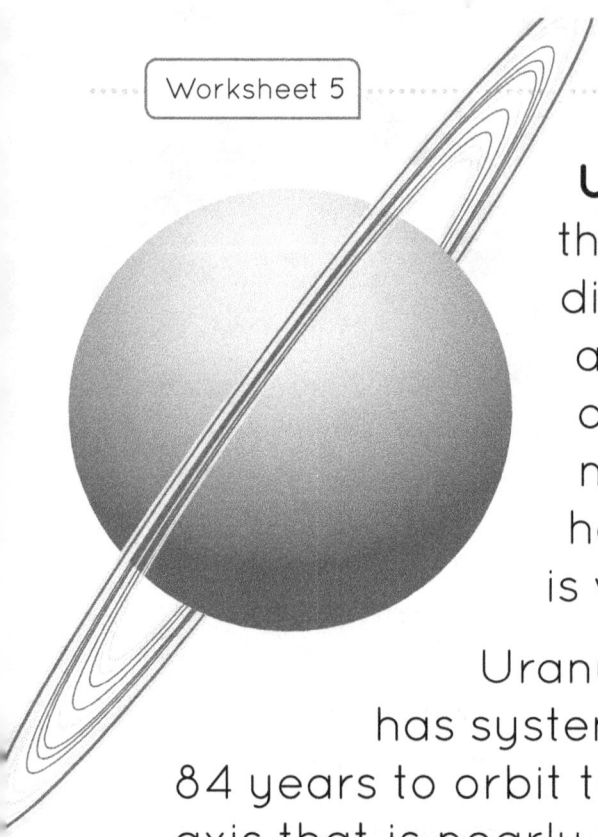

Uranus is the seventh planet from the sun. It is 51 800 kilometres in diameter. It is a cold planet, with the average temperature being -215 °C. Its atmosphere and surface are mostly made up of the gases hydrogen, helium, and methane. The methane gas is what gives Uranus its blue green color.

Uranus has 27 known moons in its orbit. It has system of 11 rings around it. It takes Uranus 84 years to orbit the sun once. Uranus rotates on an axis that is nearly horizontal, as though it has been knocked on its side. Due to this, its seasons last over 20 years long.

Neptune is the eighth planet from the sun. It is 30 800 kilometers in diameter. It is a cold planet with the average temperature being -220 °C.

Its atmosphere and surface are mostly made up of the gases hydrogen, helium, and methane.

Neptune has visible and very active weather patterns. This planet has fast moving winds of about 2000 km per hour!

Neptune has 13 known moons, and a system of 6 rings that are very faint. It takes Neptune 165 years to orbit the sun one time.

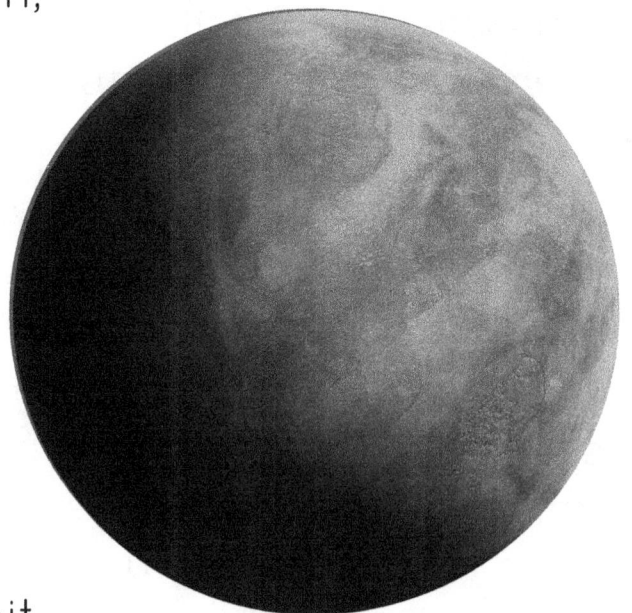

OTM2154   ISBN: 9781770789609
© On The Mark Press

Planet Trivia!

Cut out the trivia cards. Place in a pile face down. Take turns reading aloud with a partner. Each correct answer gets a point.

Which planet is mainly covered with water? A. Mars **B. Earth** C. Neptune	What is the object that all planets orbit around? **A. The sun** B. The moon C. Mercury
Which planet is nicknamed the Red Planet? A. Jupiter **B. Mars** C. Venus	Which planet has a great red spot? A. Mars B. Venus **C. Jupiter**
Which planet is the hottest? **A. Venus** B. Mars C. Mercury	How many terrestrial planets are there? **A. 4** B. 5 C. 6
Which planet has no atmosphere? A. Neptune B. Mars **C. Mercury**	What planet has the widest temperature range? A. Earth **B. Mercury** C. Saturn

OTM2154   ISBN: 9781770789609
© On The Mark Press

More Planet Trivia!

Cut the trivia cards out on the dotted lines. Place in the pile face down. Take turns asking questions to your partner.

Which planet has the only known life forms?

A. Mars

B. Jupiter

C. Earth

Which planet is surrounded by clouds containing a poisonous gas called sulfuric acid?

A. Venus

B. Neptune

C. Uranus

How many known planets are in our solar system?

A. 9

B. 8

C. 10

How many known dwarf planets are in our solar system?

A. 5

B. 6

C. 4

What planet is closest to the sun?

A. Mars

B. Venus

C. Mercury

What planet is furthest from the sun?

A. Saturn

B. Neptune

C. Uranus

How many days does it take Earth to orbit the sun?

A. 12

B. 225

C. 365

What planet has rusted iron in its dusty, rocky surface?

A. Venus

B. Mars

C. Jupiter

OTM2154   ISBN: 9781770789609
© On The Mark Press

Name:

Even More Planet Trivia!

Cut these last trivia cards out on the dotted line. Place in the pile face down. Each partner will create one to add to the game.

Which planet's surface is covered by hardened lava flows from volcanoes that erupted long ago?

   A. Mars   B. Mercury

   C. Venus

Which planet is the largest?

   A. Jupiter

   B. Saturn

   C. Uranus

Which planet is the smallest?

   A. Venus

   B. Earth

   C. Mercury

Which planet is the closest in size to Earth?

   A. Mars

   B. Venus

   C. Uranus

Which planet has seasons that last over 20 years long?

   A. Uranus

   B. Neptune

   C. Saturn

Which planet has moving wind speeds of about 2000km per hour?

   A. Jupiter   **B. Neptune**

   C. Uranus

OTM2154 ISBN: 9781770789609
© On The Mark Press

Name:

A Model of the Solar System

You have learned a lot of interesting facts about the planets. Now it is time to create a model of our solar system!

You'll need:

- 9 Styrofoam balls (4 large, 4 small, 1 extra large)
- assorted colors of paint and paint brushes
- a piece of flat Styrofoam (15cm x15cm)
- a large Styrofoam cup
- a piece of string (30cm long)

- 13 bamboo skewers
- a ruler
- scissors
- a marker
- masking tape

What to do:

1. Paint the Styrofoam balls the same colors as the planets. (The 4 small will be Mercury, Venus, Earth, and Mars. The 4 large will be Jupiter, Saturn, Uranus, and Neptune. The biggest one is the sun.)

2. Cut the bottom out of the large Styrofoam cup so that the "sun" can sit upright in it when the cup is turned upside down on a table.

3. Using the piece of string, place it around "Saturn". Mark the place on the string where it meets the end. Cut it to the length you need. (This will help you to make the rings).

4. Take the string you cut and make a circle with it on the flat piece of Styrofoam. Carefully outline it with the marker. Then draw another ring on the outside of this (about 2 cm away from the inner ring). Cut out the inner circle. Then cut out along the outside of the outer ring.

5. Once the planets are dry, use the skewers to connect them to the sun. To make them different distances from the sun, you may need to cut or break off pieces of some of the skewers.

6. Use pieces of skewers, masking tape, and a marker to create 'flag' like labels for the planets.

OTM2154 ISBN: 9781770789609
© On The Mark Press

THE VIEW FROM EARTH

LEARNING INTENTION:

Students will learn about the positions of the sun, Earth, and its moon in our solar system.

SUCCESS CRITERIA:

- recognize the positions of the sun, Earth, and its moon in our solar system
- describe how the Earth rotates on its axis and how it orbits the sun
- explain why the Earth experiences different seasons as it orbits the sun
- describe how the moon rotates and orbits the Earth, moving as one unit orbiting the sun
- demonstrate the rotation and orbit of the Earth and its moon around the sun

MATERIALS NEEDED:

- a copy of "Day and Night" Worksheet 1 for each student
- a copy of "Revolving Around the Sun" Worksheet 2 for each student
- a copy of "A Look at the Moon" Worksheet 3 for each student
- a copy of "Rotate and Orbit!" Worksheet 4 and 5 for each student
- a large ball, a small ball, a hat, 2 index cards (for each group of 3 students)
- a globe
- masking tape, markers, pencils

PROCEDURE:

*This lesson can be done as one long lesson, or be divided into two shorter lessons.

1. Using Worksheets 1, 2, and 3, do a shared reading activity with the students. This will allow for reading practice and learning how to break down word parts in order to read the larger words in the text. Along with the content, discussion of certain vocabulary words would be of benefit for students to fully understand the passage.

   Some interesting vocabulary words to focus on are:

   - axis
   - rotate
   - exposure
   - orbit
   - tilted
   - journey
   - revolution
   - hemisphere
   - equator

   (As the moon orbits the Earth, the phases change. A phase of the moon is how much of the moon appears to us on Earth to be lit up by the sun. As the moon begins to orbit the Earth, we can only see a portion of the lit up side. As it continues to orbit the Earth, we see more of the lit up side until finally the moon is on the opposite side of the Earth from the sun and we get a full moon that is 100% lit up. As the moon continues orbiting, we see less of the lit up side. When we can't see any of the lit up side, this is called a new moon. In the new moon phase, the moon is between Earth and the sun.)

2. *This activity can be done in groups, or as a whole group demonstration. Students will **need teacher guidance** to demonstrate the Earth's orbit around the sun correctly.

   Give them Worksheets 4 and 5 and the materials to create a model of the positioning of the sun, the Earth, and its moon in the solar system. Read through the materials needed and what to do sections to ensure their understanding of the task before they begin. At this point, use the globe to show students how the Earth is divided into two hemispheres, Northern Hemisphere

OTM2154   ISBN: 9781770789609
© On The Mark Press

and Southern Hemisphere. Pointing out where the students live on the globe would be of interest to them also.

Once students have positioned themselves, the teacher will help the "Earth" to orbit the sun correctly. This means that as the "Earth" is approaching the half way around mark, the teacher ensures that the "Earth" begins to face away from the sun, so that the Northern Hemisphere now becomes further from the sun while the Southern Hemisphere becomes closer to the sun.

DIFFERENTIATION:

Slower learners may benefit by having another opportunity to re-read the information on Worksheets 1, 2, and 3 in a small group with teacher support.

For enrichment, faster learners could access the internet to research some interesting facts about the sun, the Earth, or the moon. This could be later shared with the large group to promote discussion on the topic.

OTM2154   ISBN: 9781770789609
© On The Mark Press

Day and Night

You have learned that the planet Earth is third in line from the sun. In measured distance, it is about 150 million kilometres from the sun. Even from that great distance, the sun is Earth's powerful heat energy source.

So what is the Earth doing while it is soaking up the sun's heat energy? It is rotating! The Earth has an axis that runs from the North Pole to the South Pole.

As the Earth spins on its axis, from east to west, we experience day time and night time. The Earth takes 24 hours to complete a rotation.

Its rotation gives each part of the Earth a turn to be warmed by the sun. Life forms on Earth need the heat and light from the sun. If the Earth did not rotate, one half of the Earth would always be too hot to support life, and the other half would be in a deep freeze!

Axial Tilt of the Earth

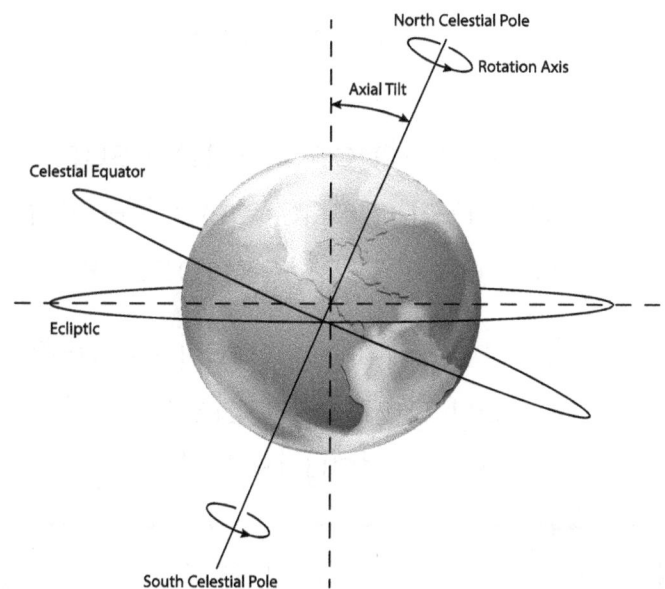

North Celestial Pole

Rotation Axis

Axial Tilt

Celestial Equator

Ecliptic

South Celestial Pole

How long does it take for the sun's heat to reach Earth? Let's do the math!

We are 150 million km away from the sun. Light travels 300 000 km per second. Dividing these numbers, it will equal to 500 seconds, or 8 minutes and 20 seconds.

OTM2154 ISBN: 9781770789609
© On The Mark Press

Revolving Around the Sun

At the same time that the Earth is rotating on its axis, it is taking a journey around the sun. This journey is called a revolution.

The Earth's journey around the sun is about 940 million kilometres long and it takes one year (365 days) to complete.

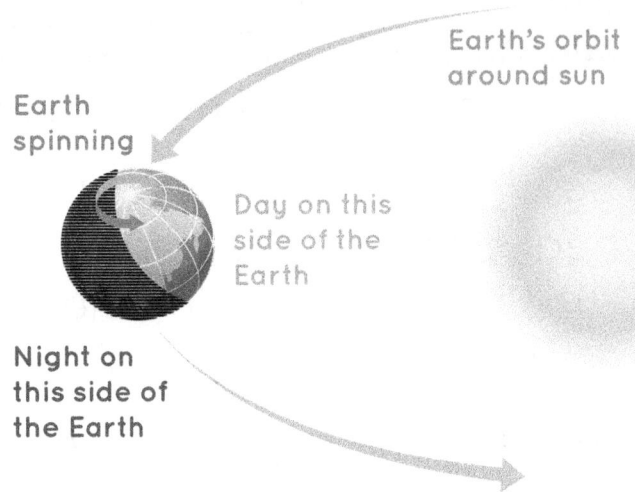

Earth's orbit around sun

Earth spinning

Day on this side of the Earth

Night on this side of the Earth

Did You Know?

The Earth is divided into two hemispheres. They are the Northern Hemisphere and the Southern Hemisphere. As the Earth orbits the sun during a year, its tilt causes the Northern and Southern hemispheres to change from more or less exposure to the sun.

As a hemisphere is tilted away from the sun, the length of a day gets shorter and it gets colder. As it is tilted toward the sun, day lengths get longer and it gets warmer. The further you are from the equator in either hemisphere, the more obvious the effect. This is why we have the different seasons in the year, Spring, Summer, Winter and Fall.

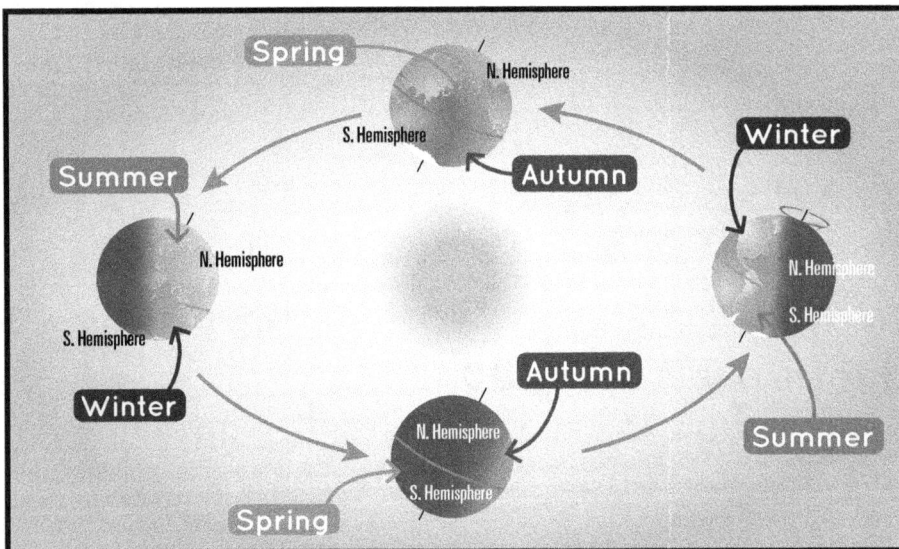

Spring — N. Hemisphere — S. Hemisphere — Autumn — Winter — Summer — N. Hemisphere — S. Hemisphere — Winter — Autumn — N. Hemisphere — S. Hemisphere — Spring — Summer

OTM2154   ISBN: 9781770789609
© On The Mark Press

A Look at the Moon

Are you wondering what the moon is doing while the Earth is revolving around the sun? It is orbiting the sun with the Earth as one unit. While it is on this journey, the moon is also revolving around the Earth, and rotating.

Have you ever wondered why the moon looks different in the sky some nights? The moon does not shine light, by itself. What we see when we look at the moon from Earth is the sun's light bouncing off the moon.

Sometimes we only see part of the moon, sometimes we see the full moon, and other times we can't see the moon at all.

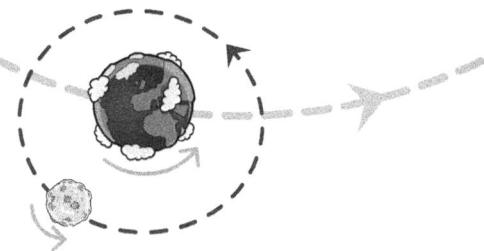

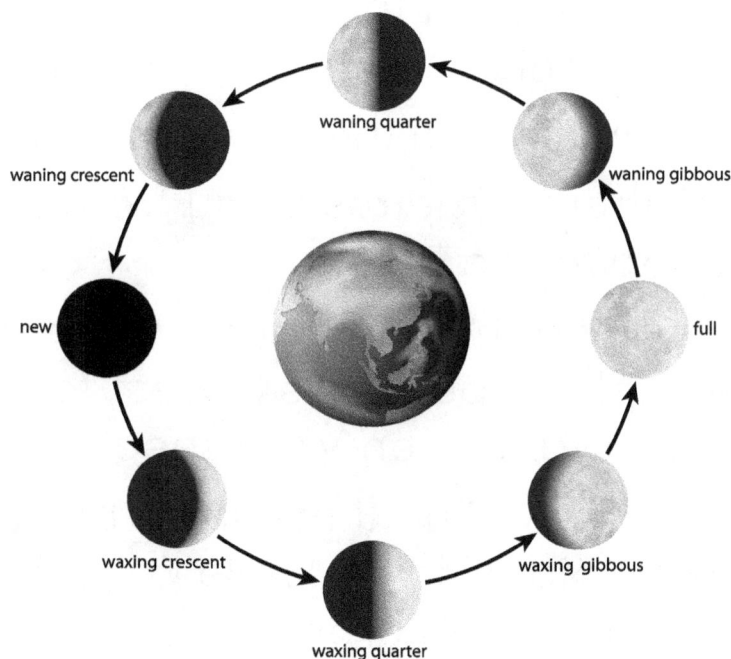

It takes about 28 days for the moon to orbit the Earth and to rotate.

The moon goes through many phases as it rotates and orbits the Earth, but from Earth we always see the same side of the moon.

waning quarter

waning crescent

waning gibbous

new

full

waxing crescent

waxing gibbous

waxing quarter

OTM2154   ISBN: 9781770789609
© On The Mark Press

Rotate and Orbit!

It is time to review the facts. You have learned that:

- Earth rotates on its axis while it orbits the sun
- because of Earth's tilt, one hemisphere is closer to the sun for half the year, while the other is farther away
- the moon rotates as it orbits the Earth
- the moon orbits the sun as one unit with the Earth

Now it is time to put these ideas into practice!

You'll need:

- a large, and a small ball
- a card with an **N** on it
- a card with an **S** on it
- a hat
- masking tape
- 2 other group members

What to do:

1. Tape the letter **N** to the hat. (**N** means Northern Hemisphere). One partner will wear the hat.

2. Tape the **S** to the lower back of this person. (**S** means Southern Hemisphere). This person is the Earth.

3. Another group member will hold the large ball (the sun).

4. The Earth will stand about 2 metres away from the sun, facing it, tilted forward a little.

5. As **your teacher** is turning the Earth around the sun, observe which hemisphere becomes closest to the sun. Record your observations on Worksheet 5.

6. The third group member will hold the small ball (the moon). The moon will rotate slowly while it orbits the Earth. Keep up with the Earth's orbit!

7. Make a conclusion about the rotation and orbit on Worksheet 5.

OTM2154 ISBN: 9781770789609
© On The Mark Press

Name:

Let's Observe

Choose words from the **Word Box** to complete your observations.

> **Northern**   **higher**   **lower**

When the Earth **started** to turn, the _____ Hemisphere was closer to the sun, and the sun stayed _____ in the sky for a longer time.

When the Earth was **half way around**, the _____ Hemisphere was farther from the sun, and the sun stayed _____ in the sky for a longer time.

Let's Conclude

Choose words from the Word Box to complete your conclusions.

> **summer**   **winter**   **shorter**   **longer**

When the sun is higher in the sky the days are _____ , just like in the _____ .

When the sun is lower in the sky the days are, _____ , just like in the _____ .

OTM2154 ISBN: 9781770789609
© On The Mark Press

CONSTELLATIONS

LEARNING INTENTION:
Students will learn about the constellations in our night sky.

SUCCESS CRITERIA:
- identify and label constellations in our night sky
- research a constellation to determine its position, special features, and story behind it
- use a Venn diagram to compare and contrast the Greek myth and Aboriginal legend of the Big Dipper's existence in the sky
- create a legend for a constellation, illustrate it in graphic text form

MATERIALS NEEDED:
- a copy of "Stars in the Night Sky" Worksheet 1 for each student
- a copy of "Constellations in the Night Sky" Worksheet 2 and 3 for each student
- a copy of "A Star Gazing Report!" Worksheet 4 and 5 for each student
- a copy of "How the Big Dipper Came to Be" Worksheet 6 for each student
- a copy of "A Legend in the Making!" Worksheet 7 and 8 for each student
- a lamp, a long extension cord
- access to computers with internet connection
- overhead projector or document camera (optional)
- clipboards, markers, pencil crayons, pencils

PROCEDURE:
*This lesson can be done as one long lesson, or be divided into four shorter lessons.**

1. Using Worksheet 1, do a shared reading activity with the students. This will allow for reading practice and learning how to break down word parts in order to read the larger words in the text. Along with the content, discussion of certain vocabulary words would be of benefit for students to fully understand the passage.

Some interesting vocabulary words to focus on are:

- galaxy
- particles
- hydrogen
- helium
- constellation
- dimmest
- kelvin
- surface
- extremely
- life span

2. Give students Worksheet 2 (or project it on a large screen using an overhead projector or document camera). Discuss the different constellations and the different positions they take in the night sky depending on the month of the year it is. To help students understand why the constellations change position, try the following exercise:

- set a lamp in the centre of an empty space (the lamp will represent the sun)
- have students stand in a large circle around the lamp, facing away from it
- tell students that their shadow is the night sky and ask them to imagine that the walls are covered in stars
- tell students to slowly walk in a circle around the lamp, and that the spot on the wall that their shadow is covering is changing as they move, just as the Earth's night sky changes as it orbits the sun

(Students may be interested to know that the stars we see in the sky depend on where in the world we are. The stars we see in the Northern Hemisphere in a certain month of the year are different from the ones that can be seen in the Southern Hemisphere in that same month of the year. We know that the seasons in the year reverse between the hemispheres, and this is the same case for the stars we are able to see in the sky.)

OTM2154   ISBN: 9781770789609
© On The Mark Press

3. Give students Worksheet 3 to complete. Encourage students to do some "star gazing" on their own one night. A suggestion is to have them draw a few that they recognize, then bring in to share with the class.

4. Give students Worksheets 4 and 5, along with a clipboard and pencil. For this activity, students will need computers and access to the internet in order to research a constellation of their choice.

5. Explain to students that many of the constellations have myths **and** legends to explain how they came to be in the sky. As an example, use Ursa Major (the Big Dipper). You can access the internet to find a Greek myth and an Aboriginal legend to share with the students. As an alternative, you can purchase National Geographic's *My First Pocket Guide Constellations* (for Greek myth), and *All the Stars in the Sky* (for Aboriginal legend) by C.J. Taylor. Students will compare and contrast the two explanations by completing Venn diagram on Worksheet 6. Follow up with a large group discussion if time permits.

6. Give students Worksheets 7 and 8. They will use the constellation that they researched in an earlier activity to create a legend of their own that explains how the constellation came to be in our night sky.

DIFFERENTIATION:

Slower learners may benefit by having a pre-determined "short list of constellations" to choose from in order to complete the research activity on Worksheets 4 and 5. Pairing up with a peer to complete this activity is another option. A further accommodation is to eliminate Worksheet 7 for these students and have them create a graphic text on Worksheet 8 of the Aboriginal legend of the Big Dipper that was shared with the entire class for the Venn diagram activity.

For enrichment, faster learners could recreate their legend using computer software called Comic Life. This is available for Mac and PC systems.

Stars in the Night Sky

Did you know that our sun is actually a star? It is just one of the many millions of stars in our galaxy!

A star is a glowing ball of dust and gases like hydrogen and helium that continues to grow as it pulls in more particles in space. The core of a star gets extremely hot because of the pressure. Some stars are brighter and hotter than others, and they come in different colors.

Stars can be blue, white, yellow, or red. The hottest and brightest stars are blue, then white ones, yellow ones are slightly cooler, and the coolest and dimmest are red. Red stars, because they are the coolest, burn very slowly and have a longer life span than blue or white stars.

Fast Fact!

A group of stars that appear to form a picture in the night sky is called a constellation. There are 88 recognized constellations that cover the entire sky.

The more mass a star has, the hotter its temperature. A star that is big, bright, and blue is extremely hot. It is called a **blue giant**.

A blue giant can range in temperature from 20 000 to 50 000 kelvin. To compare this, a yellow star like the sun has a surface temperature of about 6000 kelvin. So a blue giant star is blazing hot!

OTM2154 ISBN: 9781770789609
© On The Mark Press

Constellations in the Night Sky

There are many constellations in the night sky. We use constellations to determine an area in the celestial sky. Different constellations are visible at different times of the year. They change in position in the sky depending on the time of year it is.

Below are diagrams of what a clear night sky in the Northern Hemisphere looks like during the year.

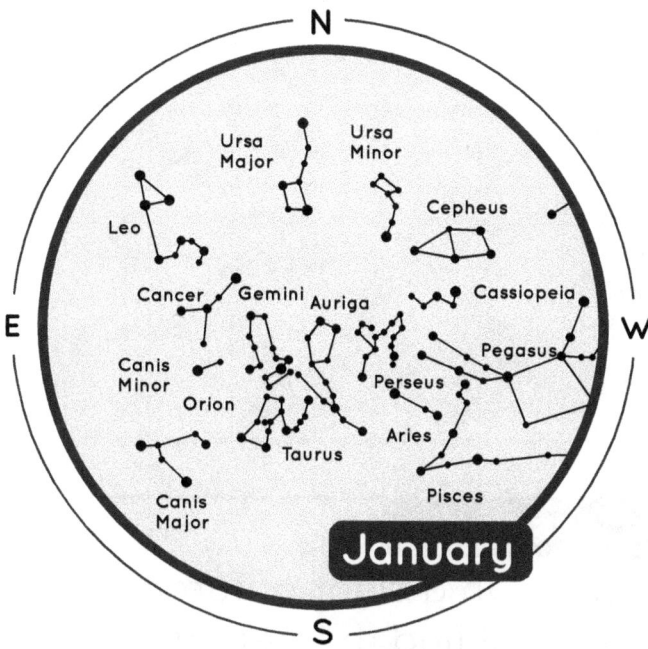

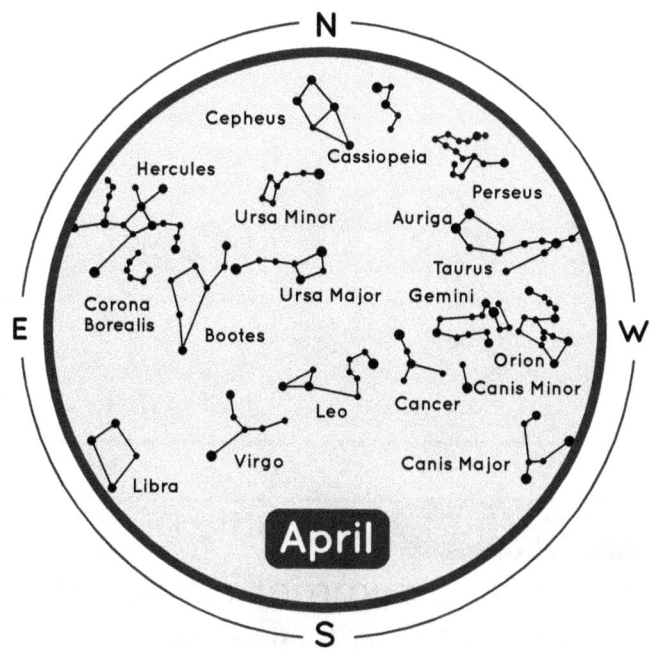

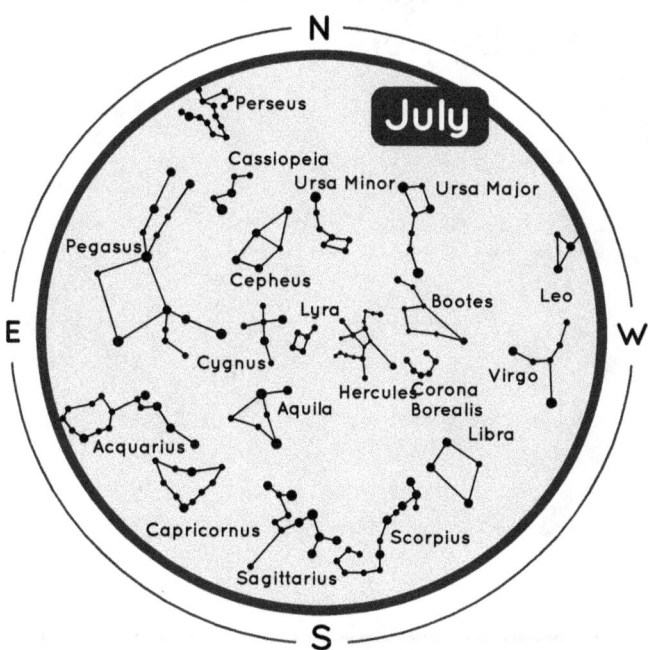

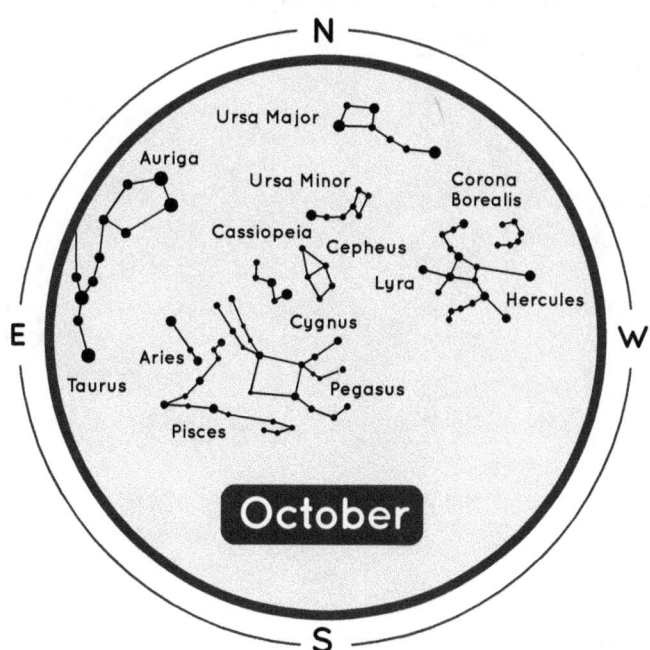

It is time to draw some constellations that are in our night sky! Use the diagrams on Worksheet 2 and the clues below to help you.

Ursa Major (the Big Dipper).

Clue: On a winter night, I sit high in the northern sky.

Orion

Clue: In the early spring time, you will find me in the western part of the sky.

Ursa Minor (the Little Dipper).

Clue: On a summer night, I am between Ursa Major and Cassiopeia.

Cassiopeia

Clue: On a clear autumn night, you will find me beneath the "Little Dipper".

OTM2154   ISBN: 9781770789609
© On The Mark Press

A Star Gazing Report!

You have learned that there are 88 recognized constellations in our night sky. Now it is time to choose one to study and write a star gazing report!

Using the internet research a constellation of your choice. Here is a graphic organizer to guide you through this task.

This is an illustration of the constellation: _____

Explain when your constellation can be seen in the night sky, including from which hemisphere.

_____

_____

_____

_____

Tell about the special features of your constellation.
For example, does it contain any special stars?

_____

_____

_____

_____

Name a few constellations that are near the
constellation that you chose.

_____

_____

_____

_____

Each constellation has its own **myth**. What is the story
behind your constellation?

_____

_____

_____

_____

_____

_____

_____

_____

OTM2154   ISBN: 9781770789609
© On The Mark Press

How the Big Dipper Came to Be

Greek mythology is only one source to help us learn about the story behind a cluster of stars in the sky. There are also **Aboriginal legends** written about how the constellations appeared in the sky.

Use the Venn diagram below to compare and contrast a Greek myth and an Aboriginal legend about how the Big Dipper came to be in our night sky.

Remember the middle of the Venn diagram is used to list similarities.

Greek myth   Aboriginal legend

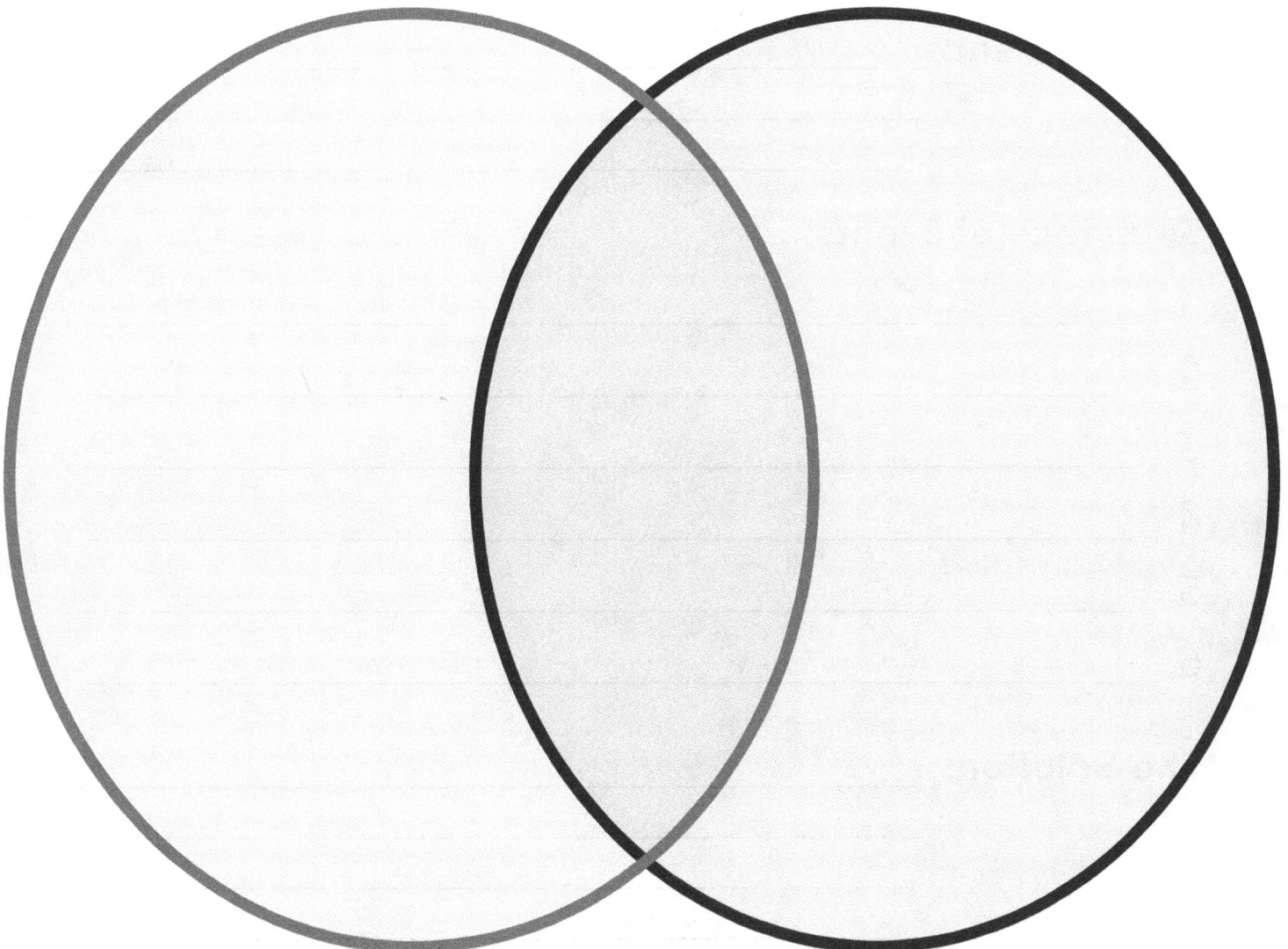

A Legend in the Making!

Use the constellation that you researched on Worksheets 4 and 5 to create a **legend of your own** to explain how it came to be in our night sky.

Use the organizer below to help you with your thinking. Then, create your legend in graphic text form on Worksheet 8.

Title: _____

Characters: _____

_____

_____

The Problem: _____

_____

_____

Sequence of Events:

1. _____

2. _____

3. _____

4. _____

5. _____

6. _____

The Solution: _____

_____

_____

_____

OTM2154   ISBN: 9781770789609
© On The Mark Press

Name:

OTM2154   ISBN: 9781770789609
© On The Mark Press